AF306384

SUSTAINABLE AVIATION FUEL AND BIOFUELS

Approach, Process Technology, Economics, Costs and Prospects

SUSTAINABLE AVIATION FUEL AND BIOFUELS

Approach, Process Technology, Economics, Costs and Prospects

Duncan Seddon

Duncan Seddon & Associates Pty. Ltd., Australia

World Scientific

NEW JERSEY · LONDON · SINGAPORE · BEIJING · SHANGHAI · HONG KONG · TAIPEI · CHENNAI · TOKYO

Published by

World Scientific Publishing Co. Pte. Ltd.

5 Toh Tuck Link, Singapore 596224

USA office: 27 Warren Street, Suite 401-402, Hackensack, NJ 07601

UK office: 57 Shelton Street, Covent Garden, London WC2H 9HE

British Library Cataloguing-in-Publication Data
A catalogue record for this book is available from the British Library.

SUSTAINABLE AVIATION FUEL AND BIOFUELS
Approach, Process Technology, Economics, Costs and Prospects

ISBN 978-981-98-1358-2 (hardcover)
ISBN 978-981-98-1359-9 (ebook for institutions)
ISBN 978-981-98-1360-5 (ebook for individuals)

For any available supplementary material, please visit
https://www.worldscientific.com/worldscibooks/10.1142/14326#t=suppl

Typeset by Stallion Press
Email: enquiries@stallionpress.com

To Judith

CAVEAT

This book is for educational purposes. The book gives overviews of available technologies for the use of renewable sources for producing biofuels with particular emphasis on producing sustainable aviation fuel (SAF). By their nature, these overviews approximate the technology and associated costs. All opinions concerning equipment manufacturers, technology, technology licensors and company strategies or services are the personal opinion of the author and may not necessarily reflect the opinions, positions or claims made by manufacturers or process licensors mentioned herein. All results are based on information available to the author at the time of writing. Changes in factors upon which the overviews are based could affect the results. Forecasts are inherently uncertain because of events and combinations of events that cannot reasonably be foreseen, including the actions of governments, individuals, third parties and competitors. No implied warranty or merchantability or fitness for a particular purpose can be given or implied by the author or publisher for any commercial decision based on the contents of this book. Any commercial decision has to be made as a consequence of considerable further work. The information, data and opinions expressed in this book may be affected as a consequence of information not in the possession of the author.

SUMMARY

Status of SAF Technology

There are a variety of routes to sustainable aviation fuel (SAF) from sugars, vegetable oils and biomass. Sugars can be used to produce ethanol (or butanol), which can be easily converted into an olefin (known technology) and thence to a paraffinic kerosene (SAF) by several known technologies.

The conversion of vegetable oils and fats to a fatty acid methyl ester (FAME) produces a fuel suitable for a diesel substitute but is generally unsuitable for SAF. Vegetable oils and fats can be converted to paraffinic blendstock by commercially known hydrogenation and isomerisation. This produces a mixture of renewable (green) gasoline, kerosene (SAF) and particularly diesel. Vegetable oil with high C_{22} and higher content can produce more jet-compatible blendstock. This process requires large quantities of hydrogen.

Biomass can be used to produce a paraffinic kerosene (SAF) by gasification followed by the commercially proven Fischer–Tropsch process or via methanol to olefins technology. Biomass can be used to produce a higher density kerosene blendstock by a series of hydrogenation–dehydrogenation processes or by pyrolysis followed by hydrogenation. These processes are under development.

These processes are, more or less, proven technologies for the production of SAF. Technologies for the production of SAF differ according to the feedstock being used. The main problem with the present known technologies is that the aviation fuel produced has too low a density for compliance with current aviation fuel standards. This means that with the present specifications for jet fuel, the SAF is used in association with higher-density petroleum-based fuels.

Higher-density kerosene for blending can be produced from sustainable sources, but, at the time of writing, the technologies are largely unproven on a commercial scale.

Cost of SAF

Routes to SAF come at a production cost higher than conventional petroleum jet fuel. This is true for all viable routes. This will require airlines, and so on, to be willing to pay a premium for SAF. Since fuel costs are a major concern for airline operations, it is not clear if SAF will be viable except in niche operations.

The main reason for the higher cost of SAF is the underlying cost of the feedstock. Agricultural feedstocks are in competition with alternative uses for the feedstocks, for instance, food, biofuels for vehicles and chemicals. This places a floor price on the feedstocks available for SAF.

Waste agricultural products, such as sawdust, do not necessarily have alternative uses. However, these are generally in short supply and may offer lower feedstock costs for niche operations producing SAF.

Production of biomass on a larger scale places the biomass in competition with sustainable power energy generating schemes, for instance the production of wood chips for large scale power generation in repurposed coal-fired generators. Again, this places a floor price on the biomass and increases the cost of the SAF.

CONTENTS

PREFACE

Most of the developed world is very concerned with the prospect of climate change and a warming planet. This phenomenon is considered to be due to rising levels of carbon dioxide in the atmosphere, which has, at the time of writing, reached over 400 ppm or 0.04%. This rise can be attributed to carbon dioxide arising from anthropogenic emissions. Much of the developed world has embraced initiatives to curb and eliminate such emissions by curbing the use of fossil fuels — coal, petroleum and natural gas — and embracing non-fossil fuel sources of energy, generally referred to as renewable or sustainable fuels.

The substitution of fossil fuels is well underway in the generation of electric power, land and sea-based transport fuels, but the substitution of fuels suitable for jet aircraft is proving a problem. This book aims at explaining the hurdles confronting jet-fuel substitution by a sustainable aviation fuel (SAF), the various approaches being trialled, and the future prospects for SAF. Because many of the routes to SAF are based on the growth and experience in the substitution of conventional petroleum fuels, (gasoline and diesel) by renewable gasoline and diesel are surveyed and discussed.

An important point of this book is that it also attempts to identify the cost of production of SAF relative to the cost of petroleum fuels.

Many of the players involved in the sustainable fuels industry are new to the technology and refining of fuels suitable for use in the automotive and aviation industries and come to the fuels business with many preconceived ideas that are unrealistic. Therefore, the first chapter deals with the current petroleum and refining industry as it applies to developments and hurdles presented to the renewable fuels sector.

Most of the approaches to producing SAF rely on agricultural products such as ethanol or vegetable oils. A generalisation is that countries that have developed agricultural industries in the production of sugar/ethanol support the expansion of that industry and favour SAF production via ethanol. This route is discussed in the second chapter. Other countries with developed vegetable oil industries are often seen to support the expansion of that industry and hence that route to SAF. The latter is particularly the case for palm oil-producing countries in the tropics. This route is discussed in the third chapter.

Countries without major agricultural industries in the sugar/ethanol or vegetable oil industries, especially in the more temperate zones of the planet, embrace the production of biomass or other organic waste streams (including municipal waste) as the source material for the production of renewable fuels and SAF. There are two approaches — gasification and pyrolysis — which are the subjects of the fourth and fifth chapters of this book.

A final chapter addresses the use of the various fuels as blendstocks to produce in part an SAF, which is consistent with the specifications for jet fuel.

A note on units: All prices are in US dollars. The oil and gas industry generally adopts US customary units with M standing for one thousand, as in MMbbl/d meaning million (thousand thousand) barrels per day. However, many projects are best expressed in SI units in which M stands for Mega or one million as in $1M, meaning one million dollars.

AUTHOR

Duncan Seddon started his career with Imperial Chemical Industries in 1973, and since that time, he has had an interest in the technology and economics of the production of large volumes of chemicals and fuels for use in the process industries. Since 1988, he has worked as an independent consultant. He has written three books:

> *"Gas Usage & Value — the Technology and Economics of Natural Gas Use in the Process Industries"*, PennWell Corporation, Tulsa, OK, 2006

> *"Petrochemical Economics — Technology Selection in a Carbon Constrained World"*, Imperial College Press, 2010.

> *"The Hydrogen Economy — Fundamentals, Technology and Economics"*, World Scientific, 2022.

With Bo Zhang, he has edited *"Hydroprocessing Catalysts and Processes — the Challenges for Biofuels Production"*, World Scientific, 2018.

He has authored over 100 publications and several patents.

1

CHALLENGES FOR RENEWABLE FUELS

Fundamentals of Crude Oil and Refinery Economics

Since the advent of the oil industry in the 1860s, the fundamental purpose of a refinery has changed very little. Its purpose is to take a liquid feedstock and convert it into a saleable product at a profit. In the 1860s, the feedstock was crude oil from Pennsylvania, and the product was kerosene for lamps. Early refinery economics was helped by the fact that the established fuels for lamps were expensive natural oils or whale oil, which sold in barrels of 40 US gallons at about US$60 each. Product marketing was helped by the invention of the distillation column, which facilitated the production of specific distillation cuts, that is, product boiling within a specific range of temperature. This improved the quality of the product, enabling the lighting kerosene to be made to a standard that produced less smoke in the lamp than competitor products, hence the name for the Standard Oil Company.

The purpose of a modern oil refinery is to convert crude oil into liquid transport fuels; these are principally gasoline, jet fuel and diesel or distillate. Every oil refinery is different, and its structure, the individual unit operations within the refinery, is dependent upon the character of the crude oil being used in the refinery and the market for the transport fuels being produced. The market is dependent on the required character of the finished fuels to be sold, which is often determined by the volume of local demand and the character of the fuel determined by local regulation.

Crude Oil

As generally understood, crude oil is a black viscous liquid produced from geological formations and traps. Chemically it is dominated by hydrocarbons of various types. It is widely available across the world, and the properties of the oil vary accordingly. Crude oil is typically produced by drilling into a porous rock layer that contains oil within its pores. The oil is kept in place by an impervious layer above it. When tapped, the oil flows to the surface by the pressure of the rock and water in the rock strata above it. Immediately above the oil layer, there is usually a layer of gas, which can also be tapped to produce natural gas. As produced, the crude oil contains dissolved natural gas, and the natural gas contains low-boiling liquids, often referred to as natural gas liquids (NGLs) or condensate.

There is thus a continuum in terms of the boiling point of hydrocarbons from methane (natural gas) to heavy viscous oils. The extremes in this range can occur. Some natural gas, such as from some fields in the Russian Arctic, contains only methane, whilst the tar sands of Canada produce solids that contain little free-running liquid. A typical refinery is constructed to use local or readily available crude oils or a mixture of crude oils and condensate. Some large refineries can economically process a wide range of crude oil, but many smaller refineries can only economically process crude oils of a specific character.

Crude Oil Character and Price

The value of a typical crude oil to a given refinery is dependent on its character. For every crude oil, this is determined by its assay, which involves distilling a sample of the oil under laboratory conditions and reporting the properties of the principal distillation cuts of interest to the refineries. Every marketed crude oil has an assay available to potential purchasers from which the purchaser can determine the value to its refining operation.

Two important properties that determine value are density and sulphur content. As well as producing noxious sulphurous gases, sulphur is inimical to the operation of modern vehicle engines, and its presence in modern transport fuel is severely restricted. Crude oil contains sulphur to varying degrees, and the amount contained is largely the consequence of the geological age of the crude oil. Oil formed from marine organisms in the earlier periods contains more sulphur than those formed in later ages; the same is true for coal measures. Because of this, oil from some parts of the world, such as the Americas, contains higher levels of sulphur than those formed in other regions, such as Southeast Asia. Crude oil of low sulphur content is easier to process than similar oil of higher sulphur content and is usually valued higher.

Density[1] is usually reported as "Degrees API". Oils with low API (e.g. below 20) are dense, whilst light crude oil and condensate have a high API (e.g. 50°). Dense, low API, crude oil has the majority of the content boiling above 360°C. This is outside the range of transport fuels, and the heavier higher-boiling material requires extensive processing to produce transport fuels from it. On the other hand, light crude oils and condensate of high API contain little material boiling in the diesel or distillate range (220°C–360°C), and so this product has to be made from other refinery inputs.

The optimum crude oils have an API in the region of 39° to 42° API and give a reasonable volume of products in the transport fuel range and smaller amounts of material boiling in the high boiling range, which require extensive processing. These intermediate crude oils are the basis of the so-called "reference crude oils" or "marker crude oils" to which many other crude oils are referenced in price. Such marker crude oils are Brent, produced in the North Sea, West Texas Intermediate (WTI), Tapis Blend (a Malaysian crude oil) and Dubai Light. These Marker Crudes are also of relatively low sulphur content.

---

[1] Density should not be confused with viscosity. Density is mass divided by volume, and viscosity is resistance to flow. Although denser crude oil is usually more viscous, this is not universally the case.

Chemical Composition of Crude Oil

Crude oil consists primarily of hydrocarbons. Some other elements, particularly sulphur, are present, but there are some, but relatively minor amounts of oxygen and nitrogen. These other elements make processing harder, and increasing amounts detract from the value of the oil. This is a particular problem with renewable oils and fuels, which often contain large quantities of oxygen. Of the hydrocarbons, unless subject to thermal treatment, olefins (alkenes) are absent or are very low in concentration in the crude oil and the distillation cuts (straight cuts) produced from it. The three dominant components of crude oil are paraffins (alkanes), *cyclo*alkanes commonly referred to as naphthenes, and aromatic compounds. Olefins enter the composition of fuels as a result of downstream processing.

Paraffins are often separated into linear and branched (*iso*-paraffins). Branched paraffins in the lighter cuts are useful for making gasoline. Some crude oils contain large amounts of long linear paraffins, which readily form wax. High wax levels lead to high pour points in the heavier distillation cuts, and the crude oil is often solid at ambient temperatures.

Naphthenes (*cyclo*alkanes) can be easily processed into aromatics, which adds value as an octane booster to the gasoline produced.

In the lower boiling ranges (below 220°C) aromatics are present as single-ring compounds with attached alkyl chains. These are effective octane boosters in the gasoline boiling range. In higher boiling ranges, polynuclear aromatics (PNAs) occur. In the extreme, these form asphaltenes that increase the density and viscosity of the crude oil making it difficult to process. PNAs are also limited to a low level in the diesel boiling range (220°C–360°C) and a high level of PNAs in crude oil can detract from its value.

In the oil assay, the paraffins, olefins, naphthene and aromatics content (PONA) are reported for the lower cuts so that the refiner of the crude can better understand the value of the distillation cut to the refining process. The wax content of crude oil is also often reported. In this regard, an oil refiner may take a differing view from a petrochemical operator because linear paraffins and wax are generally of higher value to the latter rather than the former.

Other Types of Crude Oil

As well as natural gas condensate, there are several other liquids that can be used in refineries. In many parts of the world, advanced drilling techniques are allowing the recovery of oil and gas from shale measures and tight gas sands. These are very similar, if not identical to conventional oil and gas and valued in the same way.

In other parts of the world, heavy crude oils and bitumen are being developed by mining-style operations and are being brought to market after extensive processing. This usually involves treatment to reduce viscosity, allowing the product to be pumped as conventional oil. This treatment may be the result of thermal processing, hydrogenation or mixing with lighter hydrocarbon fluids, which results in a pumpable liquid. In another process, natural gas can be converted into a crude oil-like product by the Fischer–Tropsch process. Such fluids are often referred to as synthetic crude oil and can be assayed and valued in the same manner as conventional crude oil.

Trading

Unlike most other commodities, since the oil crises of the 1970s, most crude oil is traded on the spot market. Trading takes place anywhere, but three locations dominate the trade — US Gulf, Antwerp–Rotterdam–Amsterdam (ARA) and Singapore. Spot prices at these locations are widely quoted by reporting agencies (Reuters, Platts, Petroleum Argus etc.) who publish daily bulletins for their subscribers. Other reporting agencies, such as news agencies, often take their quotations from the reporting agencies. For example, US Energy Information Agency and Organisation of Petroleum Exporting Countries (OPEC) monthly reports.

Reporting agencies make their quotations by an averaging system of active market participants, often discarding the highest and lowest quotes and volume averaging the rest. This attempts to prevent market manipulation, but this does occur in some instances, especially in the trade of lower-volume products. However, one point to note is that because the reporting agencies are independent and the method of reporting and averaging differ in detail daily (and hence monthly etc.)

averages will be different. However, rarely is the difference more than $0.5/bbl.

Figure 1.1 shows the reported values of three marker crudes — WTI, Brent and Tapis — since 1991. For most of the period, the prices move in unison. Since about 2008, with the advent of the shale oil boom in the United States, WTI has traded below the general level of Tapis and Brent. Nevertheless, the three reference crude oils have a high degree of correlation, as is shown by Figures 1.2 and 1.3.

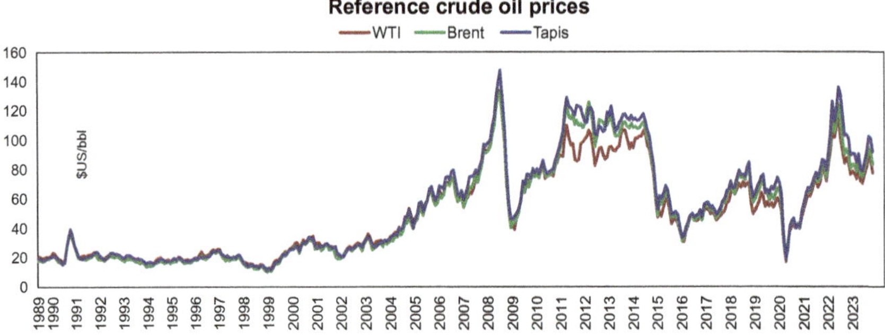

Figure 1.1.   Marker Crude Oil Prices (Singapore) from 1989 to 2022

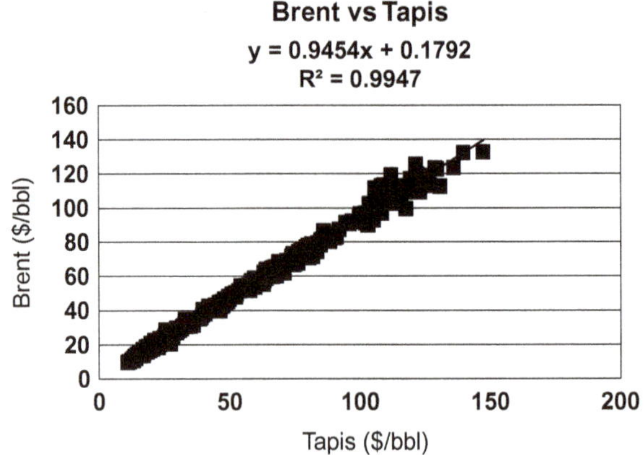

Figure 1.2.   Correlation between Brent Crude and Tapis Crude

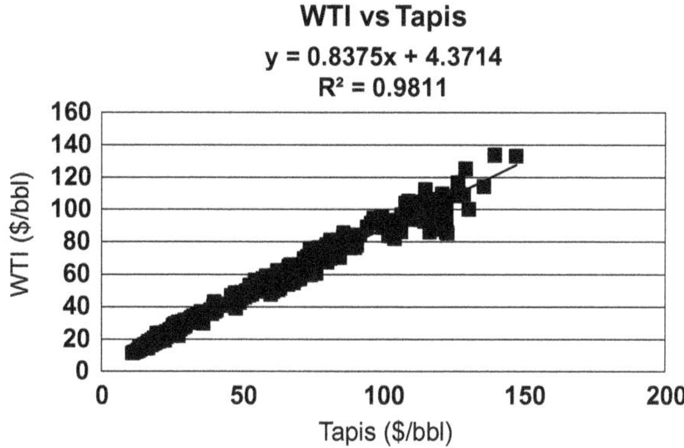

Figure 1.3.　Correlation between WTI and Tapis Crude (Singapore)

Oil Price Differentials

Distribution of Crude Oil Resources

Crude oil resources are widely distributed across the world in sedimentary basins. Many of these sedimentary basins lie in shallow water offshore. There are few countries without any sedimentary basins and hence oil or the prospect of oil discoveries. The major distributional issue is that oil demand goes hand in hand with national development, and many advanced countries require considerably more oil than indigenous supply. This is particularly the case for Western Europe, which despite major reserves in the North Sea is a major importer of crude oil and crude oil products. It is also the case with Japan, where the lack of sedimentary basins restricts significant oil production.

The history of oil in the United States has shown the country to be a major producer and exporter in the early days of the oil industry. As the national economy grew and oil production reached a plateau, the United States became an importer of oil and oil products and by the year 2000 was the world's largest importer. The shale boom from 2008 has dramatically changed this, and at the time of writing, the United States is an exporter of oil.

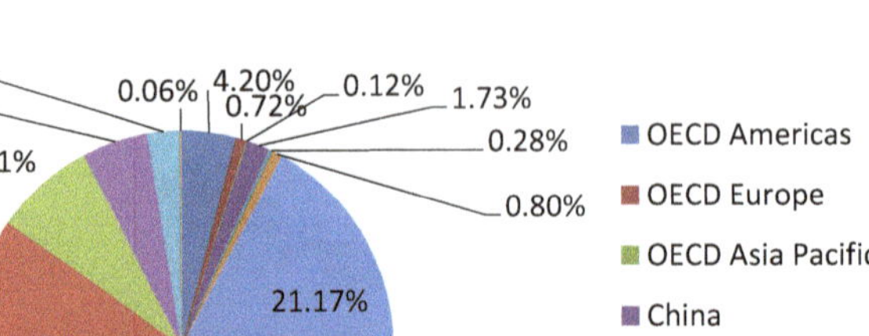

Figure 1.4. World Oil Reserves

As an aside, it is worth noting that in many parts of the world, rich nations restrict and limit the development of oil production for political purposes. For example, offshore the east coast of the United States is largely out of bounds for oil exploration, as is the northern east coast of Australia. Some areas are not explored because of the prohibitive costs, such as the Arctic (though this may be changing), and by international agreement as well as cost, such as Antarctica.

The total world oil reserves stand at over 1,500 billion barrels with the distribution indicated in Figure 1.4.[2]

The Organisation of Petroleum Exporting Countries

As many countries are set to remain oil importers, others are over-endowed with oil assets and will remain oil exporters for many years. OPEC was initially an alliance between Venezuela and Iran formed in the late 1960s. It expanded to take in other mainly Middle East oil exporters in the 1970s. OPEC countries control about 79% of the world's proven reserves (i.e. reserves that can be commercially produced) and a major portion of the export trade. OPEC flexed its power as a cartel in the early

---

[2] OPEC data.

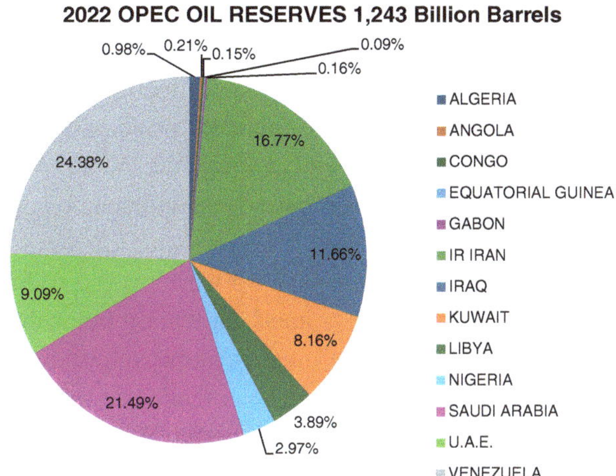

Figure 1.5. OPEC Oil Reserves

1970s by effectively determining the price at which oil would be traded. The reserves of OPEC are illustrated in Figure 1.5.[3]

How Much Oil Is There?

Peak Oil

The current crude oil demand is over 100 million barrels per day and continues to rise with increasing world gross domestic product. Despite many views to the contrary, there seems no end to the continuous rise in oil demand.

As oil demand progressively rose throughout the 1990s, the theory of peak oil came into vogue. In this theory, early work by Hubbert[4] on US oil reservoirs was used to build a picture of the ultimate reserves of oil. Back in the 1930s, Hubbert noted that for a typical oil well or field, production followed a Gaussian-like curve with production rising to a peak and then declining until the field was exhausted. In theory, by summing

---

[3] Ibid.

[4] M.K. Hubbert, *Energy Resources*, National Academy of Sciences Publication 1000-D, 1967, p. 57.

the Gaussian curves for wells, fields, regions etc., we can imagine a peak oil curve for the world. In other words, oil production would reach an inevitable peak and then go into irreversible decline. This became the subject of debate, and institutes were founded, principally using government money, to study the problem and promote the theory. At the theory's peak of interest in the 2000 to 2003 period, many became convinced that peak oil would be reached in 2004.[5]

As time went on, the critics of peak oil were proved right (at least in the short term). The critics rightly pointed out that oil fields do not decline in a Gaussian manner. As time progresses, improved and new recovery techniques cause the tail in the curve to be extended so that far more oil is recovered over time and that new oil discoveries, including synthetic oils fill the gap. The more reserves theory has been proven by the shale gale after 2008, when it was demonstrated that vast reserves of shale (note this is not oil shale, which is different) that were thought to contain little or no oil or gas could produce oil and gas using advanced techniques such as horizontal drilling and fracturing. Further improvements in the techniques have shown that oil and gas can be successfully produced in times of low oil price as well as high oil price; see Figure 1.6.

Ultimate Reserves

As geological time progresses plant and marine matter are buried within newly forming rock strata. Plant matter generally forms coal and the like, and marine matter is thought to form oil. Over the aeons, many billions of tonnes of organic matter have been buried. Table 1.1 gives data for the remaining crude oil reserves.

The data shows that there is a large portion of oil still to be produced using conventional techniques. Since this data precedes the boom in shale oil and gas production in the United States and extensive drilling of shale measures in other parts of the world, the values for remaining reserves are likely to be understated by a large margin.

---

[5]The competing approaches were debated in *Oil & Gas Journal*, July 14, 2003, with C.J. Campbell and J. Laherrere promoting peak oil theory and M.C. Lynch opposing.

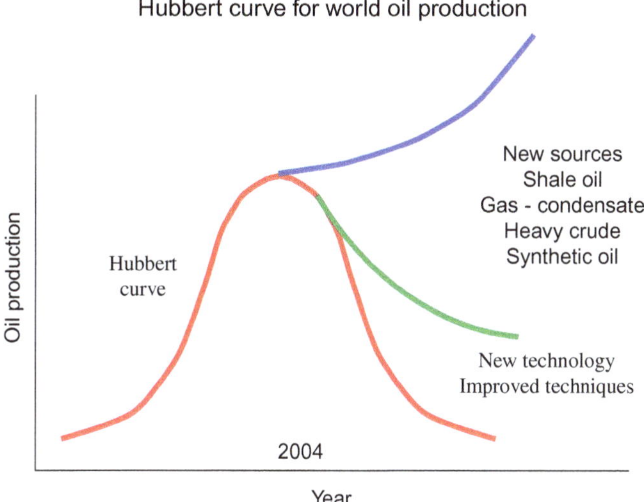

Figure 1.6.   Peak Oil

Table 1.1.   Ultimate Oil Reserves[6]

	Conventional oil (billion bbl)	Unconventional oil (billion bbl)
Remaining reserves	1,147	293
Cumulative production	1,011	7
Ultimately recoverable	2,158	300
Estimated oil in place	9,800	3,000

Furthermore, as time progresses and experience grows, more oil is recovered than initially thought possible. This is illustrated by Table 1.2 of recovery factors using various recovery techniques.

In the early days of the oil industry, recovery was confined to the first three methods, leaving most of the oil in the ground. Many of these early fields are now being reworked to extract more oil. The last group in the table is generally referred to as enhanced oil recovery (EOR) techniques and is expensive and is used when the economics merits it.

---

[6] I. Sandrea and R. Sandrea, *Oil & Gas Journal*, Global Oil reserves -I: Recovery factors leave vast target for EOR technologies, November 5, 2007.

Table 1.2. Oil Recovery Efficiencies[7]

	Percent oil in place recovered
Liquid and rock expansion	5%
Solution gas drive	20%
Gas cap expansion	30%
Gravity drainage	40%
Water influx	60%
Gas reinjection and water flooding	70%
Steam, combustion, injection of hot water, CO_2, gas, N_2, chemicals	80%

Refined Fuels and Fuel Harmonisation — Specification Fuels

Refineries

Refineries have the duty to convert crude oil into transport fuels. This is accomplished by a variety of unit process operations that are well described in the literature.[8] The following narrative focuses on the principal issues for the production and use of biofuels and concentrates on the three principal distillate fuels, which are gasoline, jet fuel and diesel. At present, other products such as aviation gasoline and marine diesel fuel play a minor role in the present story of biofuels and are not discussed.

Gasoline

Gasoline (known as petrol in some jurisdictions) is a hydrocarbon mixture with a boiling range from about 30°C to 190°C. Gasoline internal combustion engines (Otto engines) require a high-volatility fuel, so heavier hydrocarbons are not favoured. Gasoline is made by blending several different refinery streams of various compositions. In the Otto engine, fuel was evaporated on an air stream via a carburettor; in modern engines, it is

---

[7] Idem.

[8] R.E. Maples, *Petroleum Refinery Process Economics*, PennWell, 1993; J.H. Gary and G.E. Handwerk, *Petroleum Refining, Technology and Economics*, Marcel Dekker Inc., 2001.

injected into an air stream or by direct injection into the cylinder. In this manner, the cylinder contains an explosive mixture of air and hydrocarbons, which are ignited by means of a spark. The resulting explosion forces down the piston.

It is important that the air/fuel mixture does not ignite prior to the delivery of the spark. If pre-ignition occurs, it is known as an engine knock and degrades the efficiency of the process. The ability of gasoline to resist knock is measured by the octane number of the fuel, measured as research octane number (RON) for acceleration or motor octane number (MON) for steady driving.

Diesel

Diesel fuel is a hydrocarbon mixture with a boiling range from about 200°C to 360°C. Diesel engines can withstand a wider boiling range of fuel. This range is chosen primarily because low-boiling fractions have a higher value as jet fuel or gasoline, whilst heavier fuel can cause engine operational problems and increased emissions.

The crude hydrocarbon mix that goes to make diesel is often referred to as gas oil. The properties of gas oil are improved by the addition of various additives to produce the finished diesel.

Diesel fuel is used in compression ignition engines. In such an engine, a piston in a cylinder compresses air. The energy of compression heats the air in the cylinder. Fuel is injected and, after a short delay burns. This causes the pressure to rise, which forces down the piston.

Because the burning occurs over a relatively long time as a diffusion-flame front, the formation of soot from heavy molecules in the fuel is an inevitable result.[9] The formation of soot (particle matter) is one of the principal problems with the use of diesel fuel vehicles in urban environments.

Jet Fuel

Jet fuel also known as aviation kerosene, is a hydrocarbon mixture with a boiling range from about 190°C to 235°C. It thus straggles the top of the

---

[9] R. Stone, *Introduction to Internal Combustion Engines*, 3rd ed. SAE, 1999.

gasoline range and the lower fraction of the diesel range. A refinery has the flexibility to alter the cut points to maximise the jet fuel cut by incorporating the top of the gasoline range (down to about 140°C) and the bottom of the diesel range (to about 240°C) provided the final fuel remains in specification.

Fuel Specifications

Ever since the start of the automotive age, the specifications for the fuels have constantly changed. In the early days, this mainly concerned improvements to fuels to improve the performance and efficiency of internal combustion engines. Since the early 1970s, environmental concerns became the prime motivation for fuel composition changes and fuel standards. The 1970s were focused on the removal of lead (Pb) from gasoline, which led to the identification of alcohols (particularly ethanol) and ethers (particularly methyl-tertiary-butyl-ether) as alternative octane boosters to replace lead.

In the 1980s, concern focused on removing noxious gases produced from both gasoline and diesel engines. This led to the concept of fuel harmonisation in accordance with the vehicle's needs to deliver legislated environmental outcomes. An aspect of this is that a vehicle can be made anywhere, and the manufacturer could guarantee its environmental performance anywhere in the world, provided the available fuels met a specification. The specifications were determined by three parties — the vehicle manufacturers, the oil refiners and the government, who represented the public and the increasing demand for better environmental legislation. This resulted in the various major jurisdictions producing engine performance (exhaust emissions) standards that the vehicles were to deliver. As part of this process, the major vehicle manufacturers produced the Worldwide Fuel Charter (WWFC),[10] which describes the fuel quality required for various levels of performance. Tables 1.3 and 1.4 give the

---

[10]The Worldwide Fuel Charter is supported by the principal vehicle manufacturers and brings together the collected knowledge of the major automotive engine producers for the quality of gasoline and diesel. It further justifies the reasons for specifying a particular parameter by example and references.

Table 1.3. WWFC for Gasoline (Key Parameters)

Category	1	2	3	4	5	6
91RON	91.0RON, 82.0MON	91.0RON, 82.5MON	91.0RON, 82.5MON	91.0RON, 82.5MON		
95RON	95.0RON, 85.0MON	95.0RON, 85.0MON	95.0RON, 85.0MON	95.0RON, 85.0MON	95.0RON, 85.0MON	
98RON	98.0RON, 88.0MON	98.0RON, 88.0MON	98.0RON, 88.0MON	98.0RON, 88.0MON	98.0RON, 88.0MON	98.0RON, 88MON
102RON						102RON, 88MON
Sulphur (max mg/kg)	1,000	150	30	10	10	10
Oxygen (max wt.%)	2.7	2.7	2.7	2.7	2.7	**3.7**
Olefins (max vol.%)		**18.0**	10.0	10.0	10.0	10.0
Aromatics (max vol.%)	50.0	40.0	35.0	35.0	35.0	35.0
Benzene (max vol.%)	5.0	2.5	1.0	1.0	1.0	1.0
Density (min kg/m^3)	715	715	715	715	**720**	**720**
Density (min kg/m^3)	780	770	770	770	**775**	**775**
Trace elements (mg/kg)	<1	<1	<1	<1	<1	
Oxidation stability (mins)	360	480	480	480	480	480
Particulate size distribution			18/16/13	18/16/13	18/16/13	18/16/13
Fuel injector (#1, % loss)	10	5	5	5	5	5
Intake valve (ASTM D6201)		100	50	50	50	50
Combustion deposits (ASTM D6201, % fuel)		140	140	140	140	140

Table 1.4. WWFC for Diesel (Key Parameters)

Category	1	2	3	4	5
Cetane Number	48	51	53	55	55
Cetane Index	45	48	50	52	52
Density @15C (kg/m^3) min	820	820	820	820	820
Density @15C (kg/m^3) max	860	850	840	840	840
Viscosity @40C (mm^2/s)	2.0–4.5	2.0–4.0	2.0–4.0	2.0–4.0	2.0–4.0
Sulphur (mg/kg)	2,000	300	50	10	10
Trace elements (mg/kg)		<1	<1	<1	<1
Aromatics (% m/m)			20	15	15
PHA (di + tri) (% m/m)			3.0	2.0	2.0
T90		340	320	320	320
T95	370	355	340	340	340
Final boiling point (C)		365	350	350	350
Flash point (C)	55	55	55	55	55
Carbon residue (% m/m)	0.30	0.30	0.20	0.20	0.20
Water (mg/kg)	500	200	200	200	200
Oxidation stability (Delta TAN) (mgKOH/g)	0.12	0.12	0.12	0.12	0.12
FAME (% v/v)	5%	5%	5%	**5%**	Not detectable
Ethanol/methanol	Not detectable	Not detectable	Not detectable	Not detectable	Not detectable
Ash (% m/m)	0.01	0.01	0.01	0.01	0.01
Particulates (mg/kg)	10	10	10	10	10
Particulate size distribution		18/16/13	18/16/13	18/16/13	18/16/13
Lubricity (HFRR wear at 60C) (micron)	460	460	460	400	400

principal details for the gasoline and diesel. The WWFC envisages six categories for gasoline and five categories for diesel fuels:

- Category 1 fuels are for markets with no or minimal requirements for emission control; this category has now been retired for gasoline.

- Category 2 fuels are for markets with stringent requirements for emission control, such as US Tier 1, Euro-2 and Euro-3.
- Category 3 fuels are for markets with advanced requirements for emission control, such as US LEV or ULEV, Euro-4, JP 2005.
- Category 4 fuels are for markets requiring the use of sophisticated nitrogen oxides and particulate matter treatment technologies such as US Tier-2, US Tier-3, California LEV-II, Euro-4, Euro-5, Euro-6/6b, JP 2009.
- Category 5 fuels are for markets concerned with minimising carbon dioxide emissions, such as US Tier-3, California LEV-II, Euro-6c, Euro-6d, China 6a, China 6b.
- Category 6 fuels (gasoline) for markets with efficiency standards higher than Category 5. This category aims to minimise real driving emissions (RDE) required for Euro-6dTEMP. Euro-6d and China 6b.

Most countries are in at least Category 3 or higher, with advanced economies (EU, United States, Japan) in Categories 4 and 5 and moving to Category 6.

Points to note are

- The progressive removal of sulphur in both gasoline and diesel. The limits are measured at the retail pump, and since sulphur is a ubiquitous element, the 10-ppm level is to allow for some contamination from the environment.
- The progressive increase in the required octane (gasoline) and cetane (diesel).
- The high oxygen level permitted in gasoline Category 6 is to permit blends of 20% ethanol to help achieve the high-octane levels.
- The removal of oxygenates such as alcohols and fatty acid esters in the Category 5 diesel fuel.

Also of particular note are the density ranges for gasoline and diesel. Crude oil and transport fuels are among the few commodities that are universally sold on a volume basis, but it is mass that determines the mileage (distance travelled on a tank of fuel) for vehicles. The density specifications are of major interest to the trucking and goods transport industry, and the specified ranges have a pronounced impact on alternative fuels, especially biofuels.

Jet Fuel Specification

The bulk of transport fuels is for gasoline, motor diesel and marine diesel. Jet fuel also produced in refineries is about 5% of the total (about 5 million barrels/day; 5MMbbl/d).

The basic technology of modern aircraft (the jet engine) relies on the use of aviation kerosene (jet fuel) for operation. This is produced from crude oil by refining operations. In order to lower carbon emissions, there is considerable interest by the major airlines of the world in obtaining sustainable aviation fuel (SAF), which is fuel obtained with no net carbon dioxide (or methane) emissions to the atmosphere. A new aircraft produced today will still be flying in 40 years, so this time horizon must be considered for producing SAF.

An overriding objective for the airlines for an alternative fuel is that it must be a drop fuel complying in all respects to the fuel standards of the International Air Transport Authority (IATA). For effective use, such a fuel should be produced from a diverse range of sources in many parts of the world and be available to the local and international aviation industry.

The detail of the IATA specification for mainstream jet fuel is shown in Table 1.5. Jet A is mainly used in the United States and Jet A1 in the rest of the world. The fuels are interchangeable. The main differences between these two types are that Jet A1 has a lower freeze point ($-47°C$ vs. $-40°C$) and usually has static dissipater additives (SDA) added to help reduce static buildup in the fuel during flight. Jet A1 is the preferred fuel for intercontinental flights.

An SAF will have to comply with these standards and, in particular, Jet A1.

A point of note is the high energy density (especially when expressed in volumetric terms) of jet fuels. This delivers high flight mileage per fill of fuel. The high volumetric energy density of conventional jet fuel (and other transport fuels such as gasoline and diesel) is a particular feature of

Table 1.5. IATA Standards for Jet Fuel

Jet Fuel Specs			Jet A	Jet A1	ASTM test method
COMPOSITION					
Appearance			Clear & Bright	Clear & Bright	
Acidity		mgKOH/g	0.10	0.02	D3242
Aromatics	Max	vol%	25	25.0	D1319
Total Aromatics	Max	vol%		26.5	
Sulphur	Max	wt%	0.30	0.30	D1266, D1552, D2622, D4294
Sulphur mercaptan	Max	wt%	0.0030	0.0030	D3227
VOLATILITY					
IBP		C		Report	D86
10% recovery	Max	C	205	205.0	
50% recovery			Report	Report	
90% recovery			Report	Report	
Final Boiling Point	Max	C	300	300.0	
Residue	Max	vol%	1.5	1.5	
Distillation loss	Max	vol%	1.5	1.5	
Flash Point	Min	C	38	38.0	D56, D3828
Density @15C		kg/cm	775–840	775.0–840.0	D1298, D4052
FLUIDITY					
Freezing Point	Max	C	−40	−47	D2386, D5972
Viscosity @20C	Max	cSt	8.0	8.000	D445
COMBUSTION					
Lower Heating Value	Min	MJ/kg	42.8	42.80	D3338, D4529, D4809
Smoke Point	Min	Mm	25	25.0	D1322
Smoke Point AND	Min	Mm	18	19.0	D1322
Naphthalenes	Max	vol%	3.0	3.00	D1840
CORROSION					
Copper strip	Max	2h@100C	1	1	D130
THERMAL STABILITY					
JFTOT DELP @260C		mmHg	25	25	D3241
Tube Deposit	Max		<3	<3	

(Continued)

Table 1.5. *(Continued)*

Jet Fuel Specs			Jet A	Jet A1	ASTM test method
CONTAMINANTS					
Gum	Max	mg/100 mL	7	7	D381
Water reaction interface	Max		1b	1b	D1094[a]
MSEP Rating					D3948
Fuel without SDA	Min			85	
Fuel with SDA				70	
OTHER					
Conductivity					D2624
At point of use	Max		450		
At time of transfer				50–450	
BOCLE	Max	Mm		0.85	D5001

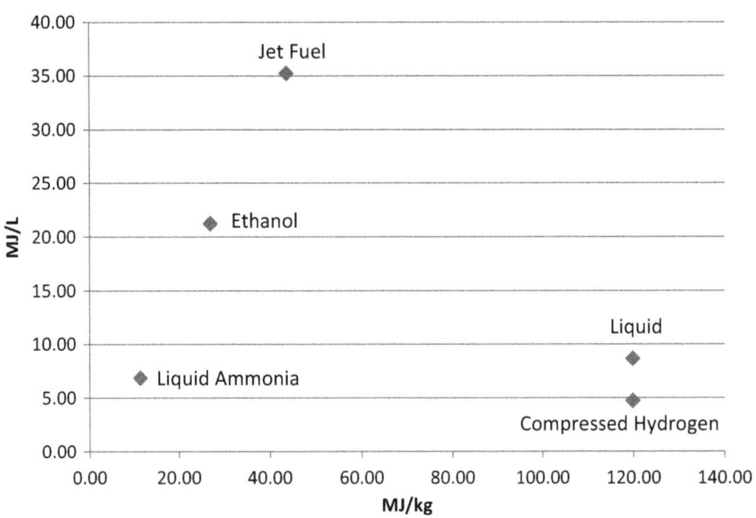

Figure 1.7. Volumetric and Mass Energy Densities of SAF Alternatives

petroleum fuels, and many of the proposed alternatives struggle to achieve reasonable volumetric energy densities.

As illustrated in Figure 1.7, hydrogen, which is proposed as an alternative to jet fuel, has a very high mass energy density but very low volumetric density. Other alternatives, such as ammonia or alcohols have lower mass and volumetric energy densities compared to conventional jet fuels, leading to lower energy per fill and hence lower range. These alternatives will struggle to achieve flying distances required in the modern aviation industry.

Considering the volatility of jet fuel (Table 1.5) places the preferred components to be hydrocarbons in the range of C_{10} to C_{15} paraffins. Furthermore, in order to meet the freezing point specification ($-47°C$ for Jet A1), these paraffins should be highly branched to achieve such low freeze points.

Prices of Refined Products

Refined products correlate directly with the prevailing crude oil price. These are widely reported by specialist reporting agencies such as Standard and Poor's Platts and Petroleum Argus. The correlation of crude oil with the three main fuels of interest for the Singapore market is illustrated in Figures 1.8 to 1.10.

Note the very high correlation coefficients (R^2) for the three correlations.

The Challenges for Renewable Fuels

The major driver for renewable fuels is the elimination of fossil fuel carbon from the fuel composition in a move by the major developed countries

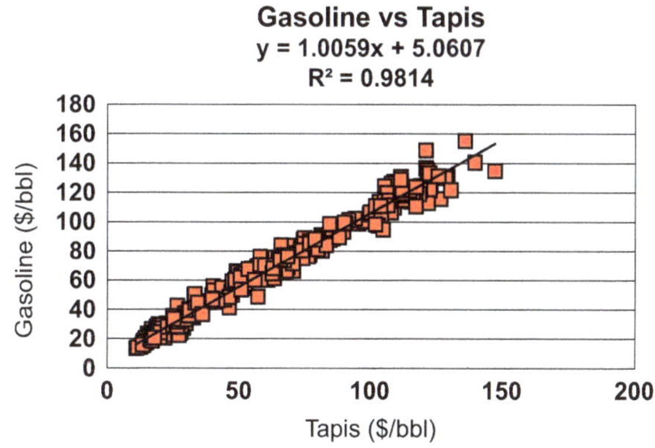

Figure 1.8. Correlation of Gasoline and Tapis Crude Oil (Singapore)

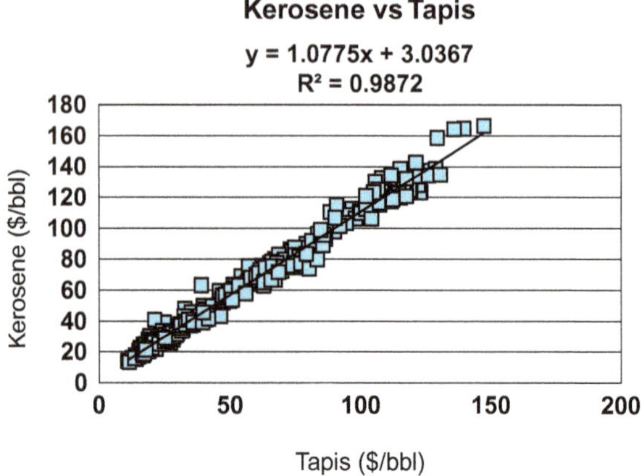

Figure 1.9.   Correlation of Kerosene (Jet-Fuel) and Tapis Crude Oil (Singapore)

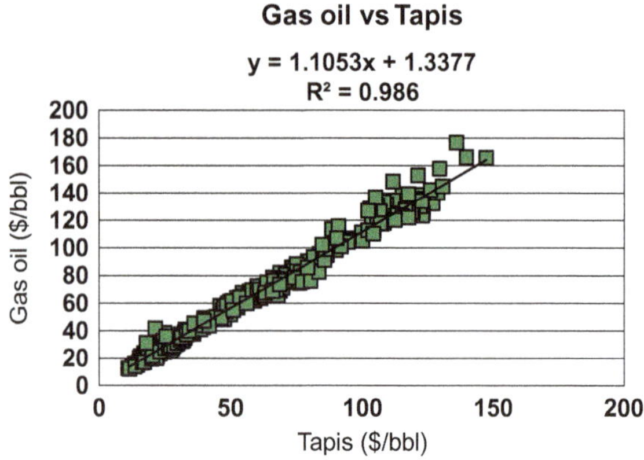

Figure 1.10.   Correlation of Gas Oil (Diesel) and Tapis Crude Oil (Singapore)

to eliminate carbon dioxide emissions from their economies; transport fuels carbon dioxide emissions are a large contributor.

The key problem for renewable fuels is that they should meet the fuel standards set out in the above tables. This is critically so for jet fuel. For gasoline and diesel, there is some allowance and governments, submitting to political pressure, often mandate the use of renewable fuels in the final fuel composition. It must be recognised that the use of renewable fuels in the mix is not favoured by oil refinery operators and, importantly, not favoured by vehicle manufacturers.[11]

Of particular interest is the concept of a "drop-in fuel". Such a fuel has properties compatible with conventional petroleum fuels and can mix in any proportion with conventional fuels and keeping the final mix within the specification.

There are, of course, other drivers for incorporating renewable fuels into the transport fuel mix, such as the disposal of agricultural surpluses, general support for farming operations and assisting a nation's economy by lowering the need to import fuel.

---

[11] Some vehicle manufacturers, for their own specific interest, build vehicles for renewable fuels, for instance, agricultural vehicles using locally produced fuel from agricultural products.

2

SAF FROM ALCOHOL-BASED FUELS

History of Use

Ethanol has been used as a transport fuel ever since the invention of the internal combustion engine. It lost its position as a preferred fuel due to the rise of low-cost petroleum fuels. After the first oil shocks of the early 1970s, there was a rise in interest in the use of ethanol, especially in countries such as Brazil, where it could be produced cheaply. During the 1980s, the use of gasoline/ethanol blends became widespread in many counties in the EU and the United States. During this period, one of the main drivers for ethanol uptake was to generate subsidies for agricultural production, especially in countries that had excess capacity to produce feedstock for ethanol production. Since 2000, the main driver has become the perceived necessity to eliminate fossil fuels from the transport sector and thereby eliminate carbon dioxide emissions to the atmosphere. This has spread the use of ethanol to other countries with the ability to produce feedstock for this end. This move to renewable ethanol eliminates the use of industrial ethanol produced by the chemicals industry from the transport fuels sector.

Ethanol

Ethanol is a simple alcohol (ethyl alcohol) that can be easily produced either by industrial processes or by biological processes such as fermentation.

As a pure chemical compound, for fuel use, the source of the ethanol is immaterial.

Substantially, most ethanol produced in the world is produced by fermentation consumed *in situ* as beverages (beer, wines etc.) and is not extracted.

Simple distillation produces ethanol containing a small amount of water. This is referred to as hydrated or aqueous ethanol or azeotrope-ethanol. There are some commercial benefits to the use of aqueous ethanol because of the savings on the process plant.

More complex distillation or molecular sieve procedures can remove the remaining water and produce anhydrous ethanol. Anhydrous ethanol usually contains some residual water, but this is limited to less than about 0.7 vol% water.

Ethanol is a pure compound with a boiling point of 78.5°C. Pure ethanol (also referred to as absolute ethanol or ethyl alcohol) is produced in small quantities for laboratory studies and the like.[1]

Commercial Grades[2]

Commercial ethanol is produced by fermentation of a suitable feedstock or by petrochemical routes. There is a wide range of suitable fermentation feedstock.

The main commercial grades are

- Neutral spirits (used for beverage fortification)
- Fuel ethanol
- Industrial ethanol, both hydrous and anhydrous

Generally, there are no official standards[3] for ethanol, rather negotiated specifications are used.

---

[1] Chemical handbooks such as Kirk Othmer's *Encyclopedia of Chemical Technology* give extensive descriptions of the properties, azeotropes, methods of manufacture etc. of ethanol.

[2] P.W. Madsson, *F.O. Lichts First World Ethanol Conference*, Mandarin Oriental, November 18–19, 1998.

[3] The American Society for Testing Materials (ASTM) has issued a standard test method for the suitability of ethanol for use in gasoline (D4806).

The specifications for anhydrous ethanol vary widely from country to country on the maximum amount of water that can be present. This is usually, though not universally, <0.1 vol% water.

Typical other specifications for anhydrous ethanol vary as well and include:

- Heads (light components)   <1 to <100 ppm
- Fusel oil[4]   <1 to <100 ppm
- Acidity   <10 to <30 ppm
- Methanol   <1 to <500 ppm
- Permanganate test   >20 to >60 minutes

Fuel ethanol will, in general, contain more water (typically up to 1.0%) and contain higher levels of impurities. However, in reviewing the various studies, many authors appear to use the term anhydrous ethanol rather than fuel ethanol for ethanol containing <0.7% water.

ASTM Standard

The American Society for Testing Materials (ASTM) has produced a specification for the use of fuel ethanol in gasoline (petrol). In this standard, the ethanol is denatured usually by the introduction of natural gasoline. Examining the standard, it is evident that the total ethanol content can be as low as 92%. The salient features of this standard are given in Table 2.1.

Table 2.1. ASTM D4806 — Fuel Ethanol Specification

Property	Limit
Ethanol (vol%)	92.1% minimum
Water (vol%)	1.0 max
Methanol (mg/L)	0.5 max
Acetic acid (wt%)	0.007 max
Chlorine (mg/L)	40 max
Copper (mg/L)	0.1 max
Denaturants (vol%)	1.96%–4.76%

---

[4]Fusel oil is a mixture of higher alcohols.

As an observation, there may be problems using such a high volume of denaturants if the ethanol is to be used in formulating a diesel blend.

Properties of Ethanol for Gasoline Blending

The main use for ethanol (and some other alcohols) is as a blendstock for gasoline. Some of the principal properties of ethanol and some other alcohols together with the alternative octane boosters methyl-*tertiary*-butyl ether (MTBE) and toluene are given in Table 2.2.

Note the generally high density of the octane boosters against the range for gasoline. MTBE is in the middle of the range. Alcohols and ethers have lower volumetric heating values than gasoline, which lowers the mileage achieved by the vehicle.

Blending Octane

Ethanol delivers a high-octane component to the gasoline. It is used in high concentrations, typically 10% to 20% in the final blend. At this level, it possibly has a minor effect as a true octane booster. This increase in octane can be used to lower the aromatics (benzene, toluene and xylenes) from the gasoline blendstock component.

Table 2.2. Some Typical Alcohol Properties

	Density (kg/L)	Oxygen (wt%)	LHV (MJ/L)	RON	MON	RVP (kPa)
Methanol	0.796	49.9	15.77	130	100	32
Ethanol	0.794	34.7	21.15	115	100	15.9
Iso-Propyl alcohol	0.789	26.6	23.79	117	100	12.4
n-Butanol	0.815	21.8	26.78	95	78	2.8
Tertiary-butyl alcohol	0.791	21.8	25.63	100	90	12.4
MTBE	0.744	18.2	28.32	110	100	54
Toluene	0.867	0.0	35.15	121	107	7.6
Gasoline[a]	0.715–0.770	~0	31.83	95	85	60

[a]WWFC Category 4

The difference between the research octane number (RON) and the motor octane number (MON) is more than 15 units for the lighter alcohols (methanol, ethanol). Vehicle manufacturers prefer this difference to be 10 or less, however, the relevance of this metric is disputed between the oil refiners and the vehicle manufacturers (note the Worldwide Fuel Charter [WWFC] difference is 10 units).

Blending ethanol (and other alcohols) is further complicated by the composition of the base gasoline blendstock as described in Table 2.3, which shows the effective blending RON and MON for three types of gasoline blendstock of similar octane values but different compositions.

Table 2.3. Blending RON and MON for Different Gasoline[5]

Gasoline blendstock	A	B	C
RON	89.2	88.4	86.1
MON	80.0	82.3	78.3
Saturates (%)	57.3	68.4	50.7
Olefins (%)	7.3	0.6	28.3
Aromatics (%)	35.5	31.0	21.0
Blending RON			
MTBE	121	120	120
Methanol	135	140	135
Ethanol	132	138	135
Tertiary-butyl alcohol	107	107	114
Toluene	115	113	114
Blending MON			
MTBE	103	109	102
Methanol	100	111	96
Ethanol	105	115	102
Tertiary-butyl alcohol	91	88	93
Toluene	92	96	94

---

[5] A.R. Braun, *Australia National Energy Research Development and Demonstration Program (NERDDP) EG/83/201.* Publishing year 1985.

The table illustrates the wide differences in observed RON and MON for ethanol as an octane booster as the gasoline composition changes. The differences are much less for MTBE and toluene.

Blending Vapour Pressure

Vapour pressure is an important parameter for the final gasoline to be delivered to the vehicle. Too high vapour pressure causes vapour lock in the fuel delivery, and this can stall the engine. It is exacerbated by hot weather, and the local limits reflect this (winter vs. summer). Refiners tend to blend butane into the gasoline to the local RVP limit. This is beneficial to the refiner as it disposes of a low-value product (butane) and decreases the density of the fuel.

Ethanol has a high blending vapour pressure, which lowers the refiner's ability to use butane in the final blend. The blending RVP of ethanol and methanol is somewhat higher than the pure component (Table 2.2) as the ethanol displaces the butane (and other light components) from the blend. Where the RVP is severely restricted (hot climates), the ethanol/blend is made by the refinery, reducing the amount of butanes and pentanes in the gasoline blendstock.

Table 2.4 illustrates the effect of the addition of alcohol to a pool (finished ready for resale) gasoline. The table clearly shows the increase in RVP on the addition of small amounts of ethanol or methanol.

Oxygen

Ethanol introduces oxygen into the gasoline (Table 2.2). At high levels, it lowers the energy density of the blend — it acts as a fuel diluent. Ethanol

Table 2.4. Effects of Addition of Alcohols into Pool Gasoline

	Gasoline	**Methanol**	**Ethanol**	**Ethanol**	**IPA**	**TBA**	**MTBE**
Volume %	100	3.0	5.0	10.0	10.0	7.0	15.0
LHV (MJ/L)	32.68	32.55	33.30	32.63	31.86	32.23	31.55
Oxygen wt%	~0	1.57	1.81	3.61	2.75	1.57	2.68
RON	91.6	92.8	92.8	93.9	93.9	92.2	94.4
MON	82.5	83	83.4	84.3	94.3	83.0	85.1
RVP (kPa)	60.0	78.2	67.8	74.6	61.7	60.5	59.1

blends (E10, 10% ethanol in gasoline) are often sold at a discount to petroleum gasoline to reflect the lower mileage delivered by an ethanol/gasoline blend. However, the oxygen has beneficial environmental properties in lowering emissions of carbon monoxide and unburnt hydrocarbons from the vehicle exhaust. This is illustrated in Figure 2.1, which shows the reduction of carbon monoxide emissions with increasing oxygen content of the gasoline.

Blending high levels of methanol and ethanol compromises the WWFC maximum oxygen content of 2.7% by weight in the gasoline. This level corresponds to a 15% MTBE blend. Ethanol blending at higher values (20%) is permitted for WWFC Category 6 fuels where high octane is required.

Ethanol and Water

Ethanol and the lower alcohols are miscible with water. Ethanol is hygroscopic (absorbs moisture from the atmosphere), and this makes the gasoline blend sensitive to water contamination. With water contamination, ethanol separates into an aqueous phase and a gasoline phase. Methanol is more sensitive than ethanol, and butanol is less. This water sensitivity precludes blending of ethanol within refinery bounds, which often has

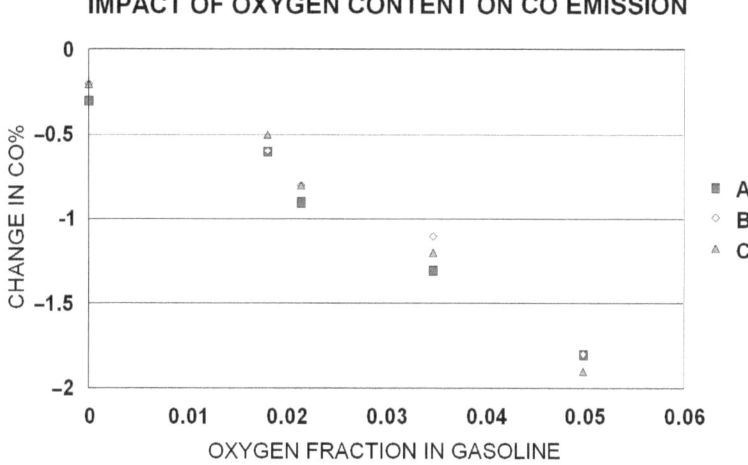

Figure 2.1. Reduction of Carbon Monoxide Emission by Increasing Oxygen Content of the Gasoline (Gasoline A, B and C are Described in Table 2.3)

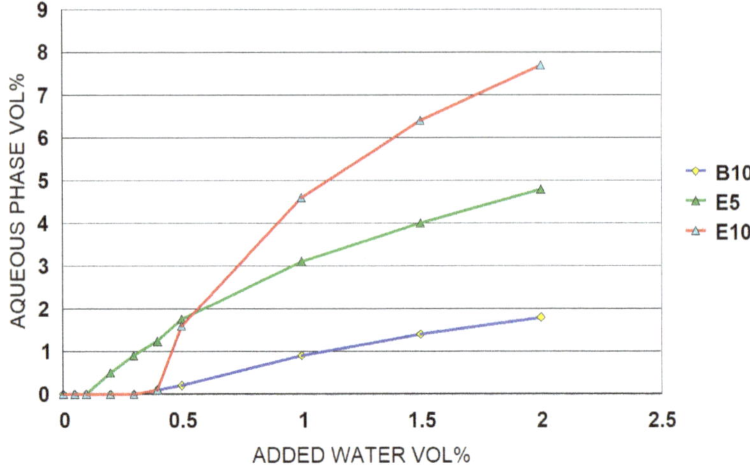

Figure 2.2. Increase in Aqueous Phase on Addition of Water to an Alcohol/Gasoline Blend (B10 is 10% Butanol, E5 is 5% Ethanol and E10 is 10% Ethanol); Adapted from BP Data

water present at the base of storage tanks and the like. This problem is solved by blending ethanol at the terminal immediately prior to dispatch to the filling station. Water contamination in storage tanks at retail outlets can induce corrosion resulting from bacterial oxidation of the ethanol to acetic acid.[6]

Water addition to alcohol/gasoline blends is illustrated in Figure 2.2. The figure shows that adding 1% water to an E10 (10% ethanol) blend results in an aqueous phase of nearly 5% as the ethanol is extracted from the gasoline by the water.

Ethanol Phase Transfer Agent

Ethanol and butanol are phase transition agents, that is, they promote the transfer of relatively insoluble hydrocarbons from the gasoline into the water phase. This means that leaks from storage tanks and pipelines of ethanol/gasoline blends can result in increased contamination of

---

[6] M. Setiyo, S. Saifudin, A.W. Jamin, R. Nugroho and D.W. Karmiadji, "Effect of Ethanol on Fuel Tank Corrosion Rate," *J. Teknol.* **80**(6), 19–25 (2018).

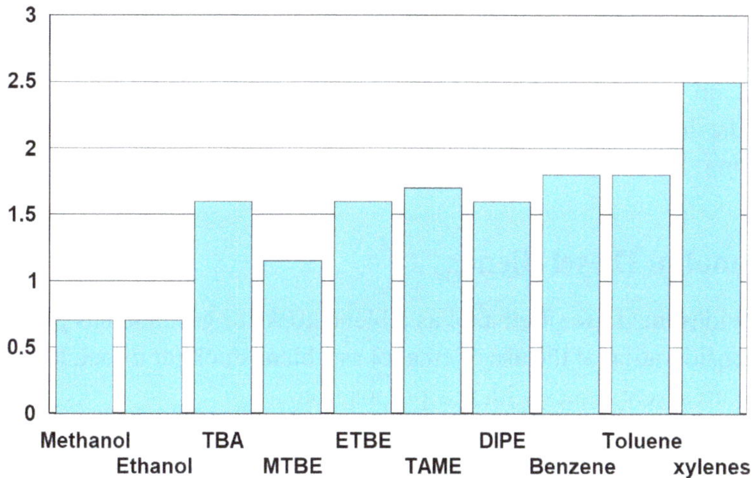

Figure 2.3. Octanol/Water Partition Coefficients for Some Gasoline Components

groundwater by aromatics in the gasoline. This property is measured by the octanol/water partition coefficient, which is illustrated in Figure 2.3.

High octanol/water coefficients indicate a propensity of the compound to be retained by soil. Low coefficients indicate materials that travel in the groundwater. The act of a phase transition agent such as ethanol is to facilitate the movement of aromatics (benzene, toluene, xylenes) in the water phase and hence movement away from a tank or pipeline leak and hence the potential to contaminate potable water supplies.

Ethanol as a Solvent

In contrast to hydrocarbon fuels, ethanol and methanol are very good solvents. The solvent effects of methylated spirit are well known to the public. This imparts solvation power to ethanol/gasoline blends and causes problems in two ways. First, it mobilises dirt and dust that collected in storage tanks and vehicle fuel tanks. This can cause blockages in fuel filters and lead to erosion in fuel pumps and fuel injectors. Second, if spilled on a vehicle Duco (vehicle paintwork), the ethanol/gasoline blend will strip the paint (especially acrylic paint) or mark the paint in the area of the spill. This is especially a problem with small spills when filling the vehicle. This is seen as a major problem from the standpoint of the vehicle manufacturers, especially when the ethanol content is higher than 10% in the blend.

The lower alcohols are absorbed by elastomers (rubbers), causing swelling. Both ethanol and butanol are similar.

Ethanol in Diesel Blends

The widespread use of ethanol as a blendstock for gasoline has prompted the consideration of the use of ethanol as a blendstock for diesel. However, there are severe issues with this proposition.

Anhydrous ethanol can be dissolved in diesel up to about 5% ethanol. The solubility is a function of temperature, and lowering the temperature results in phase separation. This commonly occurs at the cloud point of the diesel. Solubility also increases with increasing levels of aromatics in the diesel.[7]

To increase the quantum of ethanol in an ethanol/diesel blend, an additive or surfactant is required. This additive, or mixture of additives, acts as a co-solvent, maintaining the ethanol in solution, or as an emulsifier, which maintains the ethanol as an emulsion in the diesel. Ethanol–diesel blends cover a range of compositions. Ethanol can be anhydrous or hydrous. The diesel fuel can contain biodiesel fatty acid methyl ester (FAME). The additive concentration in the final blend can be several percent (e.g. 5% depending on the composition of the diesel).

Ethanol can be blended up to 20% by volume or higher in the final fuel. Most formulations concentrate on 10% or 15% ethanol blends.

There are two general approaches to formulating the fuel blend, which depend on the source of the ethanol.

Most work has been conducted on anhydrous ethanol (>99% ethanol, <1% water). This requires the addition of different agents that maintain the ethanol as a solution in the diesel.[8] Splash blending can produce this

---

[7] K.R. Gerdes and G.J. Suppes, "Miscibility of Ethanol in Diesel Fuels," *Ind. Eng. Chem. Res.* **40**, 949–956 (2001).

[8] R.L. McCormick and R. Parish, "Technical Barriers to the Use of Ethanol in Diesel Fuel," NREL/MP-540-32674. Published November 2001.

type of fuel. However, they are more susceptible to water contamination, which can break the solution and form a separate ethanol–water phase. Additional additives or dry storage and distribution systems are used to maintain the stability of the fuel.

There are several alternative approaches to the formulation of additives suitable for making ethanol–diesel blends. There are three general groups: (a) additives suitable for formulating aqueous ethanol (and methanol and possibly water) with diesel, (b) additives for formulating anhydrous ethanol with diesel, enabling splash blending and (c) additives that are aimed at producing ethanol–biodiesel–diesel blends. Within the separate groups, there is no obvious superior product. However, some formulations contain powerful solvents, which may attack vehicle parts.

If the ethanol is hydrous ethanol (95% ethanol, 5% water), the blend is formulated with additives, which retain the ethanol and water as an emulsion phase in the diesel. Emulsifier additives can improve the water tolerance of the fuels.

Adding ethanol (hydrous or anhydrous) to diesel produces a marked reduction in particulate emissions and smoke opacity. Nitrogen oxides can be simultaneously reduced. There is no apparent difference between the various technologies.

The relative quantity of unburnt hydrocarbons and carbon monoxide is dependent on the operation of the engine. However, there are many points in the engine performance characteristics where the level of carbon monoxide and hydrocarbons is lower than for conventional diesel fuel.

Ethanol diesel fuels deliver less torque (power) than conventional diesel. This results in higher fuel consumption.

Cetane

The cetane number of a diesel fuel represents a minimum delay following the injection of the fuel into the cylinder. For good operation of a large engine in a prime mover truck, the cetane number of diesel is about 45, although for smaller diesel engines, which are now widespread in the passenger vehicle fleet, the cetane number should be higher at 50 or more (see WWFC requirements in Table 2.4). Ethanol has a cetane number of only 8, so that bends require the use of cetane improvers in the mix.

Flash Point

The flash point of a liquid is the temperature at which it will be ignited by a naked flame (ignition source). Diesel has a flash point of typically 55°C or higher, which makes it inherently safe for normal ambient uses and safe in accidental spills or leaks. The flash point of ethanol is 13°C. The lowest flash point material in a blend has a disproportionate influence on the flash point of the mixture, so the flash point of ethanol/diesel blends is also about 13°C. This makes ethanol/diesel blends inherently unsafe for certain applications, such as marine diesel.

Water Content

Diesel fuel is essentially immiscible with water (to about 0.1% water). Water in diesel tanks will sink to the bottom and can be drained off. Ethanol and water are miscible in all proportions, and absolute ethanol is hygroscopic and will extract water from moist air. This results in water contamination of the diesel and easy separation of an ethanol/water phase in the tank.

Solubility

Diesel and ethanol are not good co-solvents. Under normal conditions, the concentration of ethanol in diesel is a little over 3%. Heating can dissolve more ethanol to about 8% in the final mixture. However, the addition of small amounts of water will cause an ethanol/water phase to separate. Hydrous ethanol (containing about 5% water) cannot dissolve in diesel without additives.

Vapour Pressure

Ethanol has a high vapour pressure relative to diesel fuel. The higher vapour pressure can cause vapour lock in pumping systems. However, more importantly, tank storage may become dangerous as flammable mixtures are formed in the airspace of the tank.

Ethanol/Diesel Mixtures

The overall composition of ethanol diesel blends is very similar no matter which technology is used. The composition of a 15% ethanol in a diesel blend is typically made up as follows:

- Gas oil (hydrocarbons of the appropriate boiling range); typically 81% of the final blend.
- Ethanol; 14.3% of the final blend.
- Water (up to 0.7%); because of using hydrous ethanol in the blend.
- Solvent (or surfactant and emulsifying agents) additives. This is typically a complex mixture of three or more components that perform different roles in the final product. Some of these materials have the beneficial effect of increasing the lubricity and viscosity of the final formulation, which would otherwise suffer from the introduction of ethanol. Hydrous ethanol formulations employ different types of surfactants or emulsifiers than those used for blending anhydrous ethanol. The amount of these additives can range from 1% to 5% of the final product and is typically 3%. The amount of additive is determined by the required stability, with a particular focus on the ambient temperature. In general, cold climate formulas will require a higher level of additive than summer or hot climate formulations (5% vs. perhaps only 1%).
- Another additive package that comprises pour point depressant, cold filtering improvers, injector cleaners and dyes. These additives are often polymers or large molecules. They are used in very small amounts (parts per million) but are dissolved in kerosene for ease of blending into the gas oil. They represent about 0.5% of the final diesel. To avoid confusion, it is assumed that this additive comes part and parcel with the diesel fuel, and the term additive or additive package (unless otherwise stated) will refer to the solvent or emulsifying additive package, which is necessary for forming the ethanol blend.
- Cetane improver; typically, 0.5% is added to some formulations to improve the combustion properties of the fuel.

Table 2.5. Typical Ethanol/Diesel Compositions

Diesel blendstock	81.1%
Ethanol	14.3%
Solvent	2.9%
Additives (detergents, dyes etc.)	0.5%
Cetane improvers	0.5%
Water	0.7%
TOTAL	100%

There are two general approaches to forming ethanol/diesel mixtures:

Co-Solvent Approach

This method is used for dissolving anhydrous ethanol or fuel ethanol into diesel. In the method, a solvent is added, which increases the affinity of ethanol and diesel for each other. For this, the solvent is miscible in both ethanol and diesel, and the solvent, ethanol and diesel are form a true solution or microemulsion. With this method, splash blending can be used to form the final fuel.

Water will often break these solutions, but demulsifying agents can cause the water to form a separate phase, leaving the ethanol in the diesel solution. A typical ethanol/diesel composition is shown in Table 2.5.

Emulsion Approach

This approach is suitable for hydrous ethanol. In this method, an emulsifying agent is added which forms an emulsion of the ethanol and diesel; the formation of the emulsion requires agitation; it cannot be made by splash blending. The emulsifying agent keeps microscopic globules in ethanol in suspension in the diesel. Water and methanol can also be incorporated, and hydrous ethanol can be used. The ethanol and the diesel are not in a true solution.

It is important to note that diesel fuel is a complex mixture of hydrocarbons and the actual composition depends on the refinery operations making the diesel blendstock. This means that diesel from different refineries or made to different standards has differing solvation power for ethanol or water. Diesel blendstock for making ethanol/diesel emulsions will be different in different locations and change with time. The nature of the additives for maintaining a stable emulsion may have to be changed accordingly.

Because the additive package (which is the core technology) is only a small portion of the total fuel, we would expect that variations in this component would have a secondary influence on the properties of the fuel. However, there may be some significant impacts on some specific fuel properties, such as cetane (if an improver is added) and carbon residue (if polymers are used).

The maximum level of water (even when using hydrous ethanol with 5% water) will be <1%. We would expect minimum differences between the use of fuel ethanol (<0.7% water) and hydrous ethanol (5% water) in terms of fuel performance and emissions.

Impact of Ethanol on Diesel Fuel Performance

The performance of ethanol–diesel blends is best understood from Society of Automotive Engineers (SAE) publications. SAE publications are widely regarded in the automotive industry as authoritative. The technical publications are peer-reviewed.

The importance of SAE publications is exalted when considering the automotive industries opposition to ethanol–diesel blends.

These papers are also preferred to patents and reports of demonstrations and trials because other issues are at play in these sources. In particular, as noted above, the outcome of performance and conclusions can be unduly influenced by the source of the diesel blendstock. This is often not specified in patents and rarely adequately discussed in demonstration reports. These aspects are further discussed in the following pertinent sections. The technical papers usually report the results of bench tests on engines under strictly controlled laboratory conditions. The results should therefore be repeatable by other workers in similarly equipped laboratories. This is not necessarily the case with on-road demonstrations and trials, because for no other reason, the differences in available diesel.

To ensure consistency in emissions testing, many of the laboratory studies follow agreed test procedures (protocols) involving cycling the test

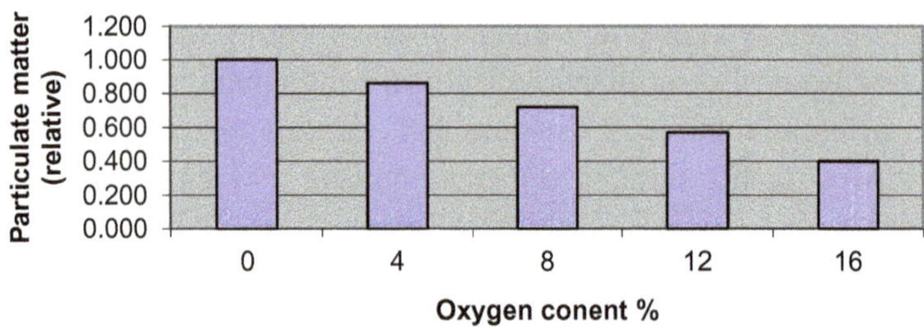

Figure 2.4.   Effect of Oxygen Content on Particulate Emissions of Diesel Fuel; after Cheng et al., SAE 2002-010-1705

engine through various modes of operation for specific time intervals. These protocols are universally agreed.

The Effect of Ethanol on Particulate Matter Emissions

Cheng et al.[9] using a 5.9 L engine illustrate the effect of oxygenates, including ethanol (anhydrous). These workers go on to show that increasing the oxygen content in a fuel (for instance, by the addition of ethanol) reduces the particulate matter (PM) emissions (Figure 2.4).

These workers also suggest that ethanol is more efficacious in reducing PM than some other oxygenates, such as dimethoxymethane (DMM).

The data in Figure 2.4 would predict that a blend of 15% ethanol (which contains about 5.2% oxygen) in diesel would reduce the PM by about 35%. This is in agreement with a compilation of private and published work by McCormick and Parish[10] on fuel–ethanol–diesel blends using different proprietary additives (Figure 2.5).

The data in Figure 2.5 was from a variety of engines and trials using anhydrous ethanol. Within the errors of such an analysis, there is no

---

[9] A.S. Cheng, R.W. Dibble and B.A. Buchholz, "The Effect of Oxygenates on Diesel Engine Particulate Matter," *SAE Technical Paper Series* No. 2002-01-1705, 2002.

[10] R.L. McCormick and R. Parish, *Advanced Petroleum Based Fuels Program and Renewable Diesel Program, Milestone Report: Technical Barriers to the Use of Ethanol in Diesel Fuel*, National Renewable Energy Laboratory, November, 2001.

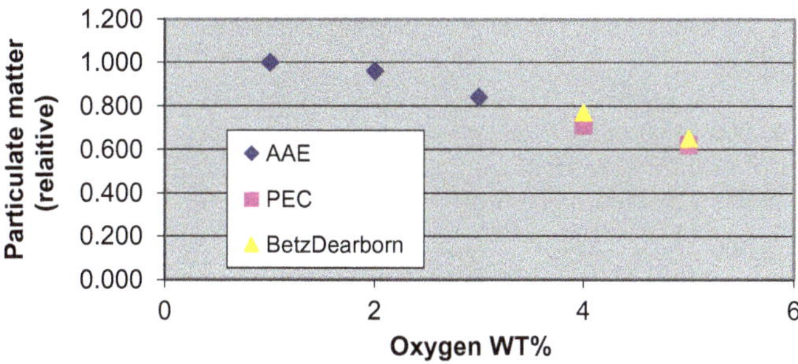

Figure 2.5. Effect of Oxygen Content on Particulate Emissions of Diesel Fuel Using Proprietary Additives; after McCormick and Parish

material difference between the various technologies. Although we might expect hydrous ethanol blends to perform similarly, there is no suitable data that could confirm this.

Effect of Ethanol on Smoke Opacity

Another study by Shih[11] compared a wide range of oxygenates, including ethanol (probably anhydrous up to 20%, with the aid of a surfactant), added to diesel fuel over a range of fuel/air ratios and engine speeds. Similar results were obtained from other workers, including a clear fall in smoke opacity with added ethanol.

The Trade-Off of Particulate (PM) and NOx Emissions

In the normal operation of a diesel engine, the conventional wisdom is that there is a trade-off between particulate and NOx emissions. This is because PM is favoured by fuel-rich combustion, whereas fuel-lean mixtures favour NOx. This is illustrated in Figure 2.6.

This illustrates that as the PM falls, there is a commensurate increase in the level of nitrogen oxide emissions and vice versa.

---

[11] L.K.L. Shih, "Comparison of the Effects of Various Fuel Additives on the Diesel Engine Emissions," *SPE Technical Paper Series* No. 982573, 1998.

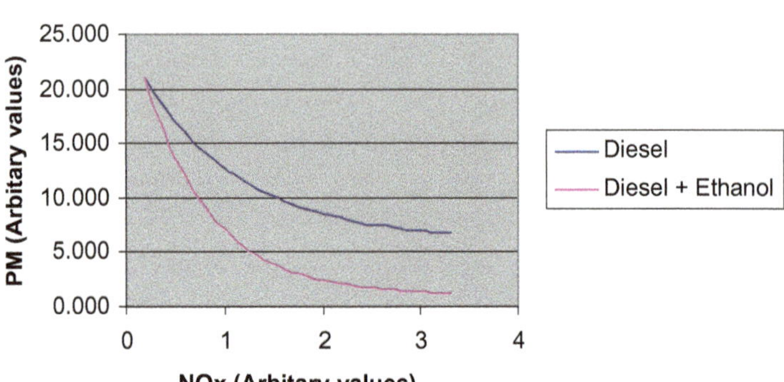

Figure 2.6. Trade-Off between Particulate and NOx Emissions of Diesel Fuel

The relative operational load on the engine complicates the emissions profile. This has been studied on a small 1.9 L engine by Cole et al.[12] These workers mapped the emissions of PM and NOx against engine torque and speed with a diesel fuel containing 0%, 10% and 15% ethanol using the Pure Energy Corporation additive package.

They showed that both PM and NOx emissions could be simultaneously reduced by the addition of ethanol over a major portion of the engine map. PM emissions increase only in the region of high speeds and light loads. This effect is illustrated in Figure 2.6 by the lower line.

For diesel engines, most of the emphasis is on the level of PM, and a modest increase in the level of nitrogen oxides has in the past been an acceptable compromise. Modern diesel engines have additional equipment to reduce particulates and NOx to the very low levels now required for operation.

Effect of Ethanol on Hydrocarbon and Carbon Monoxide Emissions

Kass et al.[13] evaluated the efficacy of 10% and 15% anhydrous ethanol (plus 2% of a blending agent provided by BetzDearborn) in a relatively

---

[12] R.L. Cole, R.B. Poola, R. Sekar, J.E. Schaus and P. Mcpartlin, "Effects of Ethanol Additives on Diesel Particulate and NOx Emissions," *SPE Technical Paper Series* No. 2001-01-1937, 2001.

[13] M.D. Kass, J.F. Thomas, J.M. Storey, N. Domingo, J. Wade and G. Kenreck, "Emissions from a 5.9 Litre Diesel Engine Fuelled with Ethanol Diesel Blends," *SPE Technical Paper Series* No. 2001-01-2019, 2001.

Table 2.6. Results of Emissions Tests Compared to Euro-3 and Euro-4 Standards[14]

Pollutant	Base diesel	10% Ethanol	15% Ethanol	Euro-3	Euro-4
CO g/kW-h	1.34	2.15	1.88	2.1	1.5
HC g/kW-h	0.27	0.55	0.47	0.66	0.46
NOx g/kW-h	5.4	5.5	5.5	5.0	3.5
PM g/kW-h	0.127	0.095	0.087	0.100	0.030

low sulphur diesel fuel (350 ppm sulphur, certified by Phillips Petroleum) in a 1999 model 5.9 L Cummins engine. They subjected the engine to an AVL 8-Mode Test Protocol.

They showed that the addition of ethanol resulted in a significant emissions *increase* in the amount of hydrocarbons and carbon monoxide over the base (0% diesel) case. Interestingly, this contrasts with the effect of the addition of ethanol to petrol, which results in the reduction of unburnt hydrocarbons and carbon monoxide.

The results indicate that 15% ethanol produced a better outcome than the fuel with 10% ethanol.

The salient details are shown in Table 2.6. In this table, Kass's emissions results have been converted to SI units and compared to the Euro-3 and Euro-4 emissions standards.

The data confirm the linear fall in PM with ethanol content. With ethanol (10% or 15%), the Euro-3 level is achieved on a 1999 model engine.

There is little change in nitrogen oxides, which remain just above the Euro-3 limit.

The higher levels of carbon monoxide and hydrocarbons are easily mitigated using commercially available oxidation catalysts and are lower than the Euro-3 standard when a 15% ethanol blend is used. The addition of ethanol failed to achieve any of the emissions required by Euro-4 regulations.

In the detailed study on the smaller engine by Cole, higher total hydrocarbon emissions were observed over most of the engine map.

---

[14] National Heritage Trust, *Setting National Fuel Quality Standards — Review of Fuel Quality Requirements for Australian Transport, Vol 1,* Coffey's Geoscience Pty. Ltd, March, 2000.

However, there were some areas where lower emissions were observed. By contrast to the Kass study, Cole showed that there were many areas of the engine map where the amount of carbon monoxide was lower than the base case.

In another SAE paper, Ahmed[15] claims the PEC additive *Puranol* reduces carbon monoxide emissions by 27% when using 15% anhydrous ethanol in diesel.

Effect of Ethanol on Engine Torque

Kass et al. measured the brake torque for each engine speed required in the AVL 8-Mode test. The results showed that overall engine speeds, the ethanol blends produced less torque than the base fuel. This reduction was about 8% for both the 10% and the 15% ethanol blends.

Cole observed a similar result in the engine mapping exercise on the smaller engine. These workers found that a 10% ethanol blend delivered 92% of the relative torque, while a 15% ethanol fuel delivered 90% of the relative torque compared to the base fuel.

Conclusion

From carefully controlled bench scale tests, the following can be concluded:

- Over most of the torque/speed curve, ethanol will reduce the PM resulting from diesel combustion and can simultaneously reduce the level of NOx emissions.
- On balance, there will be a modest rise in the level of unburnt hydro-carbons and carbon monoxide. The rise can still be within acceptable limits and can be readily eliminated by catalytic conversion.
- Ethanol (10% or 15%) reduces the torque (power) relative to conventional diesel fuel by about 8%.

---

[15] I. Ahmed, "Oxygenated Diesel: Emissions and Performance Characteristics of Ethanol-Diesel Blends in CI Engines," *SAE Technical Paper* 2001-01-2475.

Distribution and Logistics Handling

The product will have to be distributed and kept separate from conventional diesel fuels in order to prevent the contamination of conventional fuel with ethanol. Interchange of tanks should not be permitted without complete cleaning out of the tanks. This would include road haulage.

Retraining of distribution and handling personnel will be required in order that appropriate health and safety measures are to be practised.

Separation of End User Groups

It would be dangerous to allow the use of low flash point diesel by certain end-users. Of particular concern is marine diesel, which demands a high flash point product in order to prevent fires and explosions in ship or boat bilges.

Vapour Pressure

Storage

Relative to diesel fuel, ethanol has a very high vapour pressure. The major point of concern is the resultant lowering of the flash point, however, there are some other issues to be addressed. One of the general issues of concern is that the air space in tanks (both storage and vehicle tanks) can form an explosive mixture. McCormick has presented data on this issue.[16]

This is in contrast to conventional diesel, where the vapour pressure is much lower and the fuel–air mixture is below the lower-flammability limit. On the other hand, petrol has a high vapour pressure, and the fuel/air mixture in the vapour phase in a closed tank is above the upper flammability limit.

The solution to the problem may be that additional safety measures need to be added to tanks, such as the introduction of inert gas into the ullage space.

Whatever the case, the use of conventional storage tanks for the storage of ethanol–diesel blends would represent a significant change of duty

---

[16] R.L. McCormick, 7th ed. National Ethanol Conference, Technical Barriers to the Use of Ethanol in Diesel Fuel, February 12–March 1, 2002.

requiring an appropriate formal safety analysis, such as a Hazard and Operability (HAZOP) study.

Vapour Lock

Placing ethanol into diesel considerably increases the tendency to form vapour locks in vehicle fuel pumping systems. This is a consequence of the fast-acting high-pressure pumps causing cavitation in the fuel line, which then fills with ethanol vapour.

This has been noted in several trials, and there have been several proposed solutions to the problem, including the addition of coolers to cool the fuel feed lines and the addition of booster pumps within the fuel tank. This increases the cost of the vehicle.

This is a major issue for new entrants, for, although the promoters of ethanol–diesel blends aim to produce a fuel that does not require any vehicle modification, there is a view that vehicle modification (and its associated cost) is best. This highly anticipated cost acts as a barrier for some possible users.

Storage and Water Contamination

There are two issues. One is that as the temperature is lowered, ethanol–water phases may separate out from the blend. This must be overcome by precise control of the quality and quantity of the additive. This is dependent upon the water content in the blend and the quality of the diesel fuel. The following chart from Akzo Nobel data[17] (using anhydrous ethanol) illustrates the issue (Figure 2.7).

The figure clearly shows that at the 10% level, there was a significant difference between the Danish diesel (DD) and the Swedish diesel (SD) fuel.

Another issue of storage is the potential for contamination with water. Small amounts of water can get into storage tanks in the filling operations and from condensation from air in the ullage space. The potential for

---

[17] A. Lif, "Pure Alcohol Fuel and E-Diesel — Fuel Additives Technologies," Akzo Nobel Surface Chemistry AB.

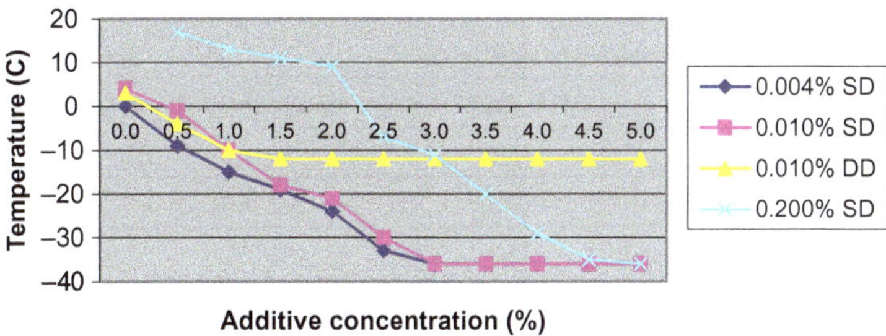

Figure 2.7. Separation Temperature for Ethanol Blended with Diesel Fuel from Sweden (SD) and Denmark (DD); after Lif

water contamination is thus dependent upon the prevailing climate, especially humidity.

Water can extract ethanol from the emulsion and form a separate phase at the bottom of the tank.

Some ethanol–diesel emulsions are said to be resistant to such contamination. These are usually the types using hydrous ethanol. Nevertheless, for widespread use, the fungibility (tendency to separate) of ethanol-diesel blends needs to be determined over a wide range of ambient conditions to ensure the product's stability in long-term storage.

The alternative approach is to move to a dry storage and handling system. This is widely practiced in the chemicals industry, where water contamination can be a problem. However, the cost of a system that completely excludes water is considerably higher than conventional tank storage for petroleum fuels.

Energy Content

Ethanol has a considerably lower energy content than diesel fuel. Typical values are shown in Table 2.7.[18]

The lower energy content translates into lower fuel economy and lower maximum power. The lower fuel economy will require some form of price advantage over conventional fuel. The power loss may be problematic for

---

[18]R.L. McCormick and R. Parish, *Advanced Petroleum Based Fuels Program and Renewable Diesel Program, Milestone Report: Technical Barriers to the Use of Ethanol in Diesel Fuel*, National Renewable Energy Laboratory, November, 2001.

Table 2.7. Energy Content (Lower Heating Value) of Typical Ethanol–Diesel Blends

Fuel	LHV (MJ/L)	% Change from diesel
Diesel	36.6	—
5% Ethanol	35.8	–2.1%
10% Ethanol	35.1	–4.2%
Ethanol	21.3	–42%

some users. This can be overcome by increasing the fuel-pumping rate but would require modification of the injector and fuel pump.

Ethanol–Kerosene Blends

Unlike gasoline and diesel ethanol blends, there is a paucity of data on kerosene (jet) ethanol fuel blends. The data presented above is used to inform the possible outcomes of adding ethanol to kerosene to form a partially sustainable aviation fuel (SAF).

Kerosene (jet fuel) has a boiling range that straddles the upper bounds of gasoline and the lower bounds of diesel. Ethanol is soluble in gasoline, but solubility in diesel is very low. Mixtures of kerosene and ethanol reflect this. Anhydrous ethanol will mix with kerosene and, in some circumstances, hydrous (95% ethanol) and kerosene can form stable mixtures. In the latter form, it is more like a stable emulsion rather than a solution.[19] The solubility of ethanol in kerosene would be expected to be influenced by the detailed composition of the kerosene. Kerosene is typically 80% to 85% paraffins and 20% to 25% aromatics. Ethanol would be expected to be more soluble in high-aromatic kerosene and less so in highly paraffinic kerosene. As will be described later in the book, many of the routes to SAF generate a highly paraffinic kerosene, which may inhibit ethanol uptake.

Partial replacement of kerosene by ethanol in gas turbine engines has been proposed[20] and the flame characteristics have been examined.[21]

---

[19] H.F. Sangian et al., "Study of Composition of the Aqueous Ethanol-Kerosene in a Single Phase," *AIP Conf. Proc.* **2694**(1), 26 (April, 2023).

[20] G. Cican and R. Mirea, "Performance and Environmental Impact of Ethanol-Kerosene Blends as Sustainable Aviation Fuels in Micro Turbo-Engines," *Int. J. Engine Res.* **25**(12), 2204–2214 (2024).

[21] J. Patra et al., *Fuel.* Studies of combustion characteristics of kerosene ethanol blends in an axi-symmetric combuster, **144**, 205–213 (2015).

From the foregoing description of gasoline and diesel ethanol mixtures, the following can be deduced:

- Adding ethanol to kerosene will reduce the heating value of the mixture. This will limit the range of the aircraft for a specific volume of fuel loaded.
- The flash point of the fuel will fall to the value for ethanol (14°C) compared to kerosene (minimum 38°C).
- There may be some stability issues at low temperatures, especially with blends made with hydrous ethanol and highly paraffinic kerosene.
- Combustion of ethanol–kerosene blends is likely to decrease carbon monoxide and PM emissions but would probably increase emissions of NOx.

The properties of ethanol prevent its inclusion in aviation fuel as a blendstock to any degree. A 5% ethanol blend may be feasible. The method of using ethanol as a basis for producing SAF is via ethylene, which is discussed later in this chapter.

World Ethanol Market

In this section, we try to analyse some of the market barriers for ethanol fuels. This involves understanding the nature of the trade and the price structure for ethanol.

World Production Capacity in Ethanol

The world's fuel ethanol production is approaching 120 billion L/year (2023).[22] The United States and Brazil dominate this production capacity (almost 80% of the total) (Figure 2.8).

Ethanol production can be sourced from either vegetable origin (e.g. fermentation plants) or from petrochemical operations (synthetic plants). Fully synthetic plants account for only about 7% of the total production. Of the vegetable sources, sugar and corn are the dominant

---

[22] Adapted from International Energy Agency data.

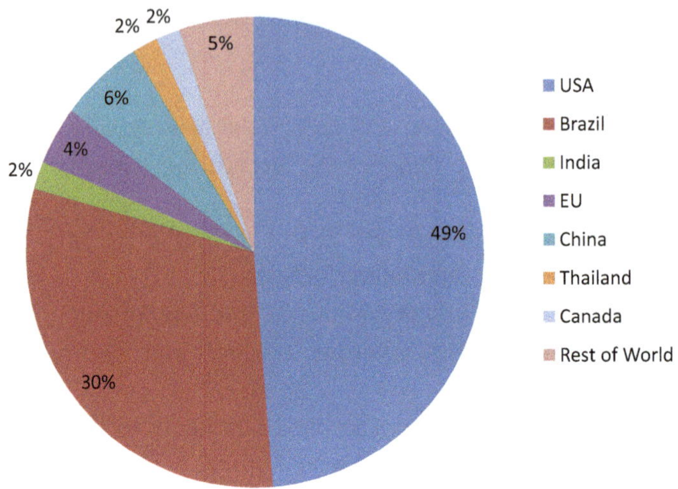

Figure 2.8. World Ethanol Production 2023

sources due to the overall dominance of Brazil and the United States, respectively.

The World Ethanol Trade

Relative to production, the trade in ethanol is very small, ranging from about 2 to 4 billion litres. The United States is the major exporter at about 0.5 billion litres. Compared to the international trade in fuel, the international trade in ethanol is almost non-existent. Nearly all production is used within specific regional markets, and there are barriers to minimise international trading.

The provision of subsidies influences and arguably (distorts) the trade. Berg[23] has estimated that only 10% of the international ethanol trade is free trade, with over 87% of the trade being influenced by subsidies and about 3% being illegal.

---

[23] C. Berg, *F.O. Lichts First World Ethanol Conference*, Mandarin Oriental, November 18–19, 1998.

Subsidies

Of the major players in the world, the majority use subsidies to promote production. Subsidisation can occur at several stages in the supply chain. These often vary on a year-by-year basis depending on the prevailing economic conditions.

Brazil is the second largest producer and consumer and has subsidised ethanol production in several ways.[24]

- Market guarantees for all ethanol produced through the national oil company PETROBRAS.
- Administered producer price by cross-subsidy from the sale of conventional gasoline.
- Administered consumer prices pegged to the pump gasoline price (ethanol sold at 59% of gasoline price, which is held at an artificially high level).
- Subsidised loans for the establishment of the agricultural and industrial production base (about 29% of total investment).

These incentives had a spectacular success, and demand for ethanol grew faster than production. This shortage, coupled with reforms in the liberalisation of the sugar market, and abnormal weather has had the consequential result that Brazil swings from importer to exporter of ethanol.

The United States is the world's largest producer and exporter. In general, producers receive a tax break. This reduces the cost of ethanol to the consumer and places it on a par with gasoline.

In normal years, the EU exports consist mainly of wine alcohol. This is a by-product of the highly subsidised EU wine industry. Wine alcohol is produced to help get rid of the EU's "wine lakes".

In addition to wine ethanol, bioethanol is subsidised by tax breaks in various EU countries.

---

[24] C. Berg, *F.O. Lichts World Ethanol 2000*, Langham Hilton, November 9–10, 2000.

Trading Logistics

Ethanol is traded in a similar manner to many other chemicals. Parcel sizes in chemical tankers are typically 5,000 to 20,000 t. A 10,000 t parcel costs about US$40/t to ship across the Pacific.

Ethanol Production Costs

Because of the existence of a wide variety of subsidies in the production of ethanol, including low-interest loans and capital grants, it is very difficult to independently assess the true production cost of fuel ethanol.

There are some data in the US market that would help an understanding of the likely production costs of ethanol, or more precisely enable limits to be identified.

However, there has been a recent consolidation of the US industry. This is resulted in the supply being controlled by a few large companies. The top four companies produce over 75% of the US supply.[25]

The Price of Ethanol in the US Market

Ethanol in the United States is a well-traded commodity, and price data is available on Chicago Board of Trade (CBOT) websites and from reporting agencies. The US Grains Council[26] reports various prices for ethanol in comparison to gasoline and octane boosters such as MTBE, toluene and mixed xylenes (Table 2.8).

The table illustrates the various price differentials as of January 18, 2023. At this time, Brazilian ethanol was selling at a marginally higher price than US Gulf ethanol. Note the high differential ($0.075/L) between anhydrous and hydrous ethanol. Other octane boosters (MTBE, toluene, xylenes) cost more on a volume basis than ethanol, which was at a slightly lower cost than gasoline at the time. The website gives further details of price history and price spreads for ethanol.

Recent price history is illustrated in Figure 2.9.[27]

---

[25] Data from Statista, *Empowering People with Data,* (2024). www.statista.com.

[26] US Grains Council, *Developing Markets. Enabling Trade. Improving Lives*, https://grains.org, 2025.

[27] *Data Taken from Trading Economics Graphs.* http://tradingeconomics.com, 2025.

Table 2.8. Ethanol Pricing for January 18, 2023 Illustrating Price Differentials ($/L)

	Current price ($/L)	**Previous year ($/L)**
Ethanol (FOB, US Gulf)	0.610	0.599
Anhydrous (FOB, Santos, Brazil)	0.619	0.716
Hydrous (FOB, Santos, Brazil)	0.544	0.628
MTBE (FOB, US Gulf)	0.805	0.669
Toluene (FOB, US Gulf)	0.947	0.780
Mixed xylenes (FOB, US Gulf)	0.925	0.787
Gasoline (FOB, US Gulf)	0.633	0.600

Figure 2.9. Recent Price History for Ethanol in the United States

From this, we can make the deduction that the production cost is likely to be no higher than the lowest delivery price (23c/L) plus the tax subsidy. This estimates the production cost to be in the region of 40c/L ($506/t).

In support of this, Energy Securities[28] have performed a detailed analysis of the production costs across all the states in the United States. Notional production costs range from about 35.9c/L for a corn-belt state such as Minnesota to 43.6c/L in Oregon.

---

[28] Data from Energy Security Analysis Inc. in *Supply and Cost Alternatives to MTBE in Gasoline* California Energy Commission, 1998, Technical Appendices, list major producing plants.

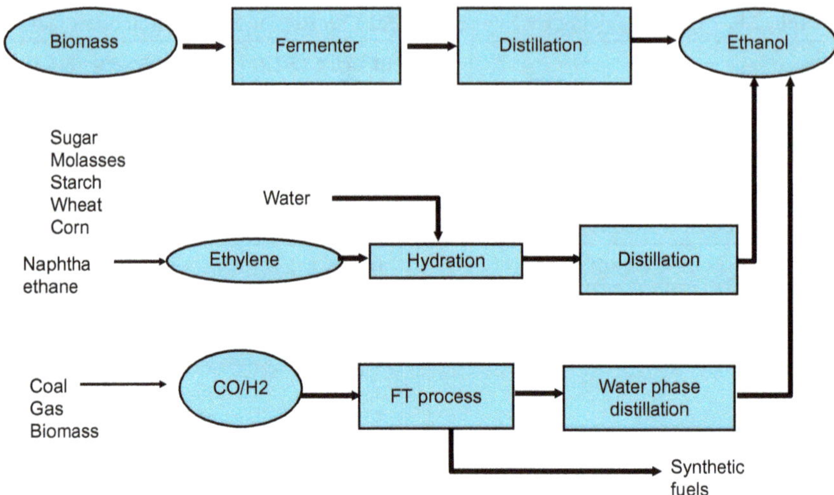

Figure 2.10.   Routes to Ethanol

This gives a useful basis for the development of production costs for ethanol fuel blends since, broadly, production costs using sugar would be expected to be like the production cost of corn.

Ethanol Production Technology and Production Costs

There are several alternative routes to the production of ethanol to be considered, figuratively illustrated in Figure 2.10. Three routes are considered: (i) fermentation routes, (ii) hydration of ethylene and (iii) via synthesis gas (carbon monoxide and hydrogen, CO/H_2).

For biofuels, the main method is by fermentation routes, which can be supplemented by other routes that do not involve carbon emissions.

Fermentation Routes

The fermentation routes require an agricultural crop that will provide sugar or cellulose. The main production methods involve the use of sugar (sugar cane from tropical and subtropical areas, sugar beet in temperate areas) and corn syrup (mainly in the United States). Some other feedstocks are used in special circumstances, such as excess wine. Because of the volumes of ethanol required and the inability of conventional agriculture to provide the volume of feedstock, there has been an interest in the fermentation of cellulose materials, particularly lignocellulose from wood waste and the like.

The basic reaction from glucose is

$$C_6H_{12}O_6 = 2C_2H_5OH + 2CO_2$$

This equation teaches that 180 parts by weight of sugar (considered as glucose) will give 92 parts of ethanol and 88 parts of carbon dioxide.

Ethanol Production from Sugars

The steps in the process are (i) agricultural production of the biomass, (ii) production of an appropriate sugar syrup (e.g. molasses or corn syrup), (iii) fermentation of the syrup with added enzymes (yeasts), (iv) separation of the ethanol liquor and distillation to produce a hydrated ethanol and (v) further distillation or other methods to produce anhydrous or fuel ethanol.

Variations on this process using different enzymes and genetically modified enzymes can give butanol as the principal product. An early version of this, known as ABE fermentation[29] was used prior to the advent of the petrochemicals industry to produce acetone, n-butanol and ethanol using *Clostridium acetobutylicum*. Many variants have been developed. *Clostridium butyricum* gives iso-propanol rather than acetone, *Clostridium aurantibutyricum* gives acetone, iso-propanol and butanol, *Clostridium tetanomorphum* gives ethanol and butanol.

A simplified flow diagram for ABE fermentation is shown in Figure 2.11. Batch processes of 700 kL (200,000 gal.) over 40 to 60 hours were conducted in two phases (i) to produce acids and (ii) to produce solvents. Hydrogen and carbon dioxide were produced as by-products. The final solvent concentration was in the range of 18 to 22 g/L. Solid by-product was used as cattle food. The route involved several different enzymes producing several intermediates in the fermentation process.

---

[29] First reported by Pasteur in 1861.

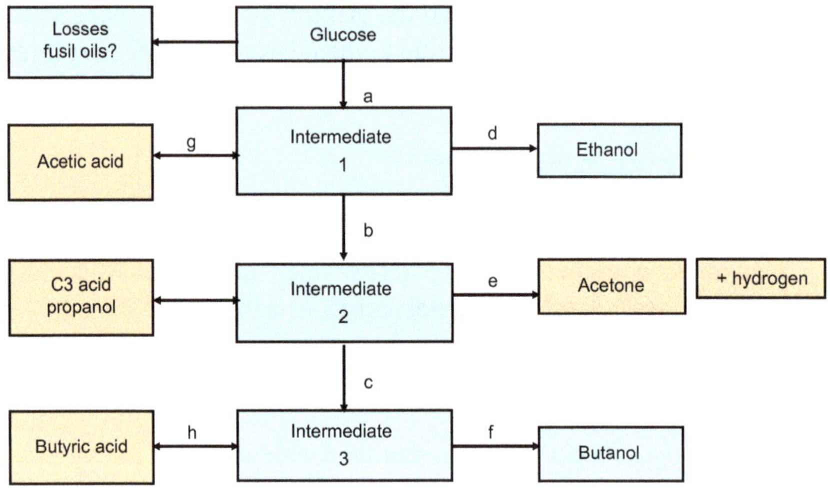

Figure 2.11. Schematic of ABE Fermentation

Glucose is fermented to an intermediate (a) which produces C_2 products, principally ethanol (d) with acetic acid as a by-product (g) and some losses to fusel oils (higher alcohols).

$$C_6H_{12}O_6 + H_2O = 2C_2H_5OH + 2CO_2$$

$$C_6H_{12}O_6 + 2H_2O = 2CH_3CO_2H + 2CO_2$$

The intermediate produces a second intermediate (b), which produces C_3 products, principally acetone (and hydrogen) with C_3 acids and propanol as by-products.

$$C_6H_{12}O_6 + H_2O = 2CH_3COCH_3 + 4H_2 + 3CO_2$$

This intermediate forms a third intermediate (c), which produces C4 products, principally butanol (f) with butyric acid (h) as a by-product.

$$C_6H_{12}O_6 = 2C_4H_9OH + H_2O + 2CO_2$$

$$C_6H_{12}O_6 = 2C_3H_7CO_2H + 2H_2 + 2CO_2$$

Assuming all acids are ultimately reduced to the corresponding alcohol, the approximate yields are shown in Table 2.9.

Table 2.9. Approximate Yields for ABE Fermentation[30]

	Mol. Wt.	Moles	Mass	Yields (wt%)
Glucose	180	1	180	
Acetone	58	0.22	12.76	7.1%
Ethanol	46	0.21	9.66	5.4%
Butanol	74	0.6	44.4	24.7%
Hydrogen	2	0.99	1.98	1.1%
CO_2	44	2.22	97.68	54.3%
TOTAL			166.48	
Losses			13.52	7.5%
Mass of solvents				37.1%

Table 2.10. Approximate Yield for Butanol Production

	Mol. Wt.	Moles	Mass	Yields (wt%)
Glucose	180	1	180	
Butanol	74	0.8	59.2	32.9%
Hydrogen	2	1	2	1.1%
CO_2	44	2	88	48.9%
Water	18	1	18	10.0%
Losses				7.1%

In recent times, these processes have been modified to produce butanol products using genetically modified enzymes. Several practical issues are that butanol is toxic to enzymes, and this limits the butanol concentration in the brew (limit typically 2%) and limits the production rate (4.5 g/L/h) over immobilised bacteria and enzymes. The high extraction cost can be minimised by solvent extraction rather than distillation. Allowing for some losses (like ABE), an approximate yield is given in Table 2.10.

---

[30] Adapted from D.T. Jones and D.R. Woods, "Acetone-Butanol Fermentation Revisited," *Microbiol. Rev.* **50**(4), 484–542 (December, 1986); Table 1.

Using the yield in the above data and published data[31] the outline production economics are given in Table 2.11; see Appendix for a discussion on methodology. The original capital cost data has been updated to 2022. In general, the feedstock for sugar-based plants is molasses. The price of molasses ranges widely relative to sugar, and to simplify the approach, raw sugar is assumed to be the principal feedstock, which typically has a value of 86% of the prevailing price for refined sugar (assumed $500/t in Table 2.11). The original work estimated the cost of a lignin feedstock, and this has been costed at $15/t to the facility. The work has been expanded to add a case for butanol; the economics of the ethanol plant has been taken as a guide with some added costs for enzymes and a lower plant efficiency.

The sensitivity to the prevailing feedstock price is shown in Figure 2.12.

Table 2.11. Estimates of Cost of Production for Ethanol and Butanol

		OAHU	**MAUI**	**KAUAI**	**Butanol**
Feedstock		Lignin	Raw sugar	Raw sugar	Raw sugar
Capacity (Gal/year)		15,000,000	15,000,000	10,000,000	10,000,000
ML/y		56.79	56.79	37.86	37.86
CAPEX (2022)	$	103,136,784	77,667,665	58,061,821	77,667,665
Return on capital % Capex for (2-year construction; 20-year life; 10% Dcf; 2% royalty)		13.84%	13.84%	13.84%	13.84%
Return on capital (ROC)	$/y	14,274,131	10,749,205	8,035,756	10,749,205
CAPEX ($/L)	$/L	0.251	0.189	0.212	0.284
Working capital	$	400,000	291,000	217,000	291,000
Return on working capital (10%)	$/y	40,000	29,100	21,700	29,100
OPEX					
Enzymes	$/y	1,454,400	0	0	727,200
Other chemicals	$/y	1,115,329	1,154,286	769,524	1,115,329

---

[31] M. Yancey, *Economic Impact Assessment for Ethanol Production and Use in Hawaii: An Interim Report*, Hawaii Ethanol Workshop, November 14, 2002.

Table 2.11. (Continued)

		OAHU	MAUI	KAUAI	Butanol
Fuel oil	$/y	2,980,950	2,833,333	1,888,889	2,980,950
Electricity	$/y	2,040,000	1,165,714	777,143	2,040,000
Denaturant	$/y	655,714	655,714	437,143	0
Other costs	$/y	484,757	196,856	137,585	196,856
Labour and overheads	$/y	1,059,537	1,059,537	753,729	1,059,537
Administration	$/y	2,777,196	2,387,153	1,910,661	2,387,153
Employees		31	31	22	31
Operating costs	$/y	14,933,119	11,659,312	8,148,120	13,531,625
	$/L	0.263	0.205	0.215	0.357
Feedstock					
Production (ML/a)	ML/y	56.79	56.79	37.86	37.86
	t/y	44,824.347	44,824.35	29,882.9	30,666.6
Plant efficiency		60%	90%	90%	75%
Feedstock (as raw sugar)		138,858.03	92,572.02	61,714.68	75,999.83
Price (refined sugar $500/t)	$/t	15	432.1	432.13	432.13
Feedstock costs	$/y	2,082,870.5	40,003,422	26,668,948	32,842,034
	$/L	0.0367	0.704	0.704	0.868
Total costs ($/a)	$/y	31,290,120	62,411,939	42,852,824	57,122,864
$/L	$/L	0.551	1.099	1.132	1.508
	$/t	698.06	1,392.4	1,434.0	1,862.7
	$/gal	2.085	4.160	4.284	5.879

Production from Corn

Horstrand[32] of Iowa State University has developed a model for estimating the production cost of ethanol produced by corn fermentation, considering various by-products from the process such as corn oil and distillers grains. This work is updated and reproduced on the Illinois Farmdoc website.[33]

---

[32] D. Hofstrand, *Tracking Ethanol Profitability / Ag Decision Maker*, iastate.edu. www.extension.iastate.edu/agdm/energy/htm/d1-10.html: downloaded 2025.

[33] farmdoc daily, https://farmdocdaily.illinois.edu/2023/03: downloaded August 28, 2023.

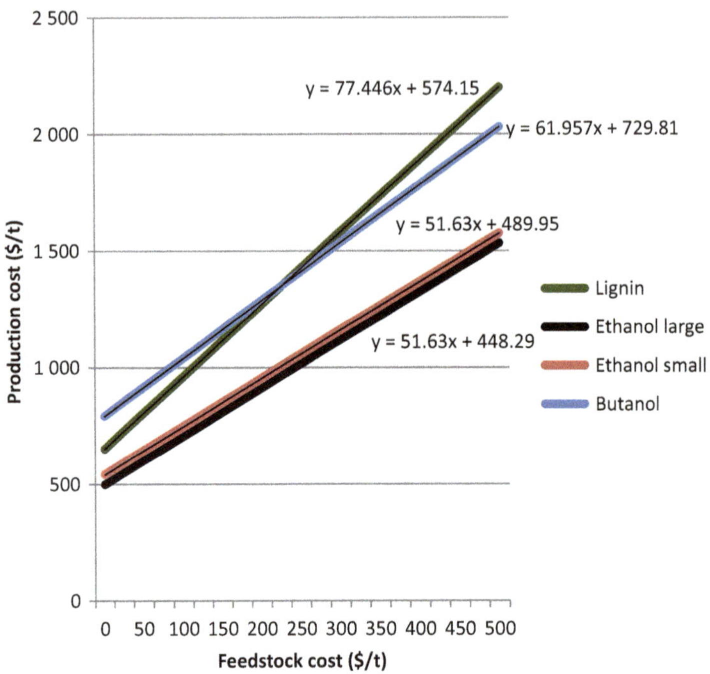

Figure 2.12.   Sensitivity of Ethanol Production Cost by Fermentation to Feedstock Cost

The base data has been transformed to the methodology used in this work and reproduced in Table 2.12.

The estimate of $0.66/L ($861/t) should be compared to that for a similar sized plant in Table 2.11 using the higher-cost raw sugar as feedstock with an estimated cost of production of $1.13/L. If sugar or molasses is available at $250/t, then the cost of production of the sugar route falls to $0.83/L, similar to that estimated for the corn feedstock case.

Ethanol Production by Lignocellulose (Corn Stover)

Table 2.11 gave estimates to produce ethanol from lignin. A more detailed production cost estimate is available from Aden et al.[34] In this work some

---

[34] A. Aden, M. Ruth, K. Ibsen, J. Jechura, K. Neeves, J. Sheehan, B. Wallace, L. Montague, A. Slayton and J. Lucas, "Lignocellulosic Biomass to Ethanol Process Design and Economics Utilizing Co-Current Dilute Acid Prehydrolysis and Enzymatic Hydrolysis for Corn Stover" NREL/TP-510-324328, June 2002.

Table 2.12. Production Cost from Corn (Data from Illinois Farmdoc daily)

Capacity	Mgal/y	100
Capital cost	$/gal	2.11
2023	$/gal	3.89717
Non-feed costs	$/gal	0.26
Distiller dried grains	lb/bsl	16.00
	lb/lb	0.286
Corn oil	lb/bsl	0.90
	lb/lb	0.0161
Distiller dried grains	$/ton	225
Corn oil	$/gal	0.55
	$/L	0.1453
Capacity	Mgal/y	100
	ML/y	378.5
	kt/y	298.64
CAPEX (2022)	$M	389.717
ROC (2-year construction, 10%dcf, 20-year life)	% Capex	13.57%
	$M/y	52.88
Working capital (30 days)	Mgal	8.333
Notional price	$/gal	2
	$M	16.67
10% return		1.67
Feedstock	kt	876.15
Notional price	$/t	250
	$M/y	219.04
Gas Costs	GJ/t	3.60
Notional price	$/GJ	4
	$M/y	12.61
Non-feed operating cost	$/gal	0.26
	$M/y	26.00

(*Continued*)

Table 2.12.   (*Continued*)

Gross costs	$M/y	312.20
By-products		
DDGS	$M/y	62.07
Corn oil	$M/y	0.013
Net costs	$M/y	250.11
Ethanol	$/gal	2.501
	$/L	0.661

Table 2.13.   Estimates for the Cost of Corn Stover ($/t)

Cost of stover	% of total	2002
Bale and stage	47%	$29.14
Transport	23%	$14.26
Farmer premium	18%	$11.16
Fertilizer	12%	$7.44
TOTAL		$62.00

attention was paid to the cost of the biomass corn stover which is summarised in Table 2.13.

The 2002 cost distribution was made up of collecting and baling the stover, transport to the facility, a contribution to the farming operation and the cost of fertiliser. The then estimate of $62/t was considered high and it was assumed that optimisation of the logistics would reduce this to $30/t. Escalating this value to 2022 using cost indices would deliver a very high cost of around $100/t. This again intuitively seems too high, and a cost of $50/t is used in this analysis.

The authors also discussed the impact of transport distance on the haulage costs, which are illustrated in Figure 2.13.

The figure illustrates the increasing cost of haulage as the distance for the transport of stover increases. From this data, the inference is that distances beyond 80 miles (about 130 km) would considerably increase the cost of the feedstock. This distance seems about the optimum for biomass-based operations. This is an important parameter, as it serves to limit the scale of the operation and so deny full economy of scale to a project.

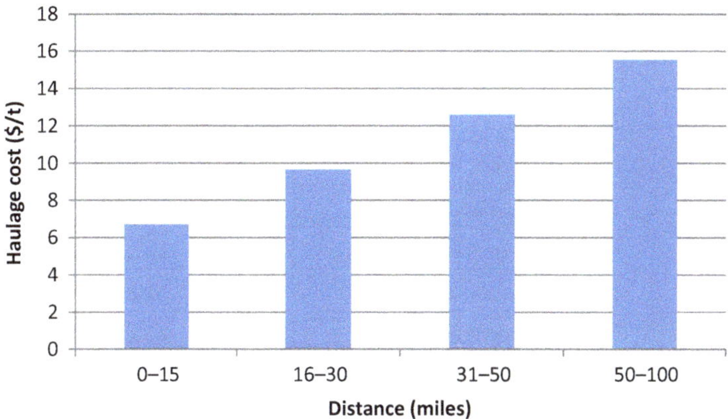

Figure 2.13. Impact of Distance on Haulage Costs for Corn Stover

The economics are developed for producing 201,000 t/y ethanol from a facility using 800,000 t/y of corn stover. The breakdown of the capital and operating costs is given in Table 2.14, which gives the 2000 costs and costs escalated to 2022.

The cost estimate for the production cost for ethanol from such a facility is developed in Table 2.15. The estimate is for a production cost of $769/t; the figure for a year 2000 facility was $433/t. The 2022 figure is very similar to the value developed for the production cost in Table 2.11 of $700/t.

Ethanol by Ethylene Hydration

The current global ethylene capacity is 224 Mt/y (2022) with an annualised growth of 6%/y. The major source of ethylene is by thermal steam cracking of naphtha or ethane (gas oil and LPG can also be used). Most ethylene is used to make polyethylene and copolymers with other olefins. Other uses include the production of industrial ethanol, but this has declined in recent years due to the availability of large volumes of fermentation ethanol.

Ethanol for purely industrial uses (e.g. as a solvent) is made by hydration of the ethylene.

$$C_2H_4 + H_2O = C_2H_5OH$$

Table 2.14. Capital and Operating Cost of a Lignocellulose Facility for the Production of Ethanol

	2000	2022
CAPEX	$M	
Feed handling	$7.50	$18.92
Pre-treatment	$19.00	$47.92
Neutralisation	$7.80	$19.67
Fermentation	$9.40	$23.71
Distillation	$21.80	$54.98
Wastewater	$3.30	$8.32
Storage	$2.00	$5.04
Boiler	$38.30	$96.59
Utilities	$4.70	$11.85
EQUIPMENT (ISBL)	$113.80	$287.00
Off-sites	$1.70	$4.29
Site	$5.90	$14.88
Field costs	$24.30	$61.28
Construction	$30.30	$76.42
Contingency	$3.60	$9.08
Other (permits, start-up)	$17.90	$45.14
Subtotal	$83.70	$211.09
TOTAL CAPEX	$197.50	$498.10
Operating costs (OPEX)	$M/y	$M/y
Stover ($30/t; $50/t)	$23.17	$38.62
Enzyme	$6.98	$17.60
Other	$7.80	$19.67
Salaries and wages	$2.10	$5.30
Overhead	$1.29	$3.26
Maintenance	$2.27	$5.73
Insurance	$1.82	$4.59
TOTAL OPEX	$45.44	$94.77
By-products		
Electricity (18.747 MW)	$6.43	$7.84

Table 2.15. Estimate of Cost of Production for Ethanol from Corn Stover

Ethanol	kt/y	201.44
CAPEX	$M	498.10
ROC (2, 10, 20)[a]	% Capex	13.57%
ROC (2, 10, 20)[a]	$M/y	67.59
Working cap (30 days, 300/t)	$M/y	0.53
Capital charges	$M/y	68.12
Stover	$M/y	38.62
Enzyme etc.	$M/y	37.28
Feedstock	$M/y	75.89
OPEX	$M/y	18.88
Subtotal costs	$M/y	162.89
Byproduct credits	$M/y	7.84
Net cost	$M/y	155.05
Unit production cost	$/t	769.73
Original (2000)		433.20

[a]Return on capital for a 2-year construction, 10%Dcf, 20-year facility life

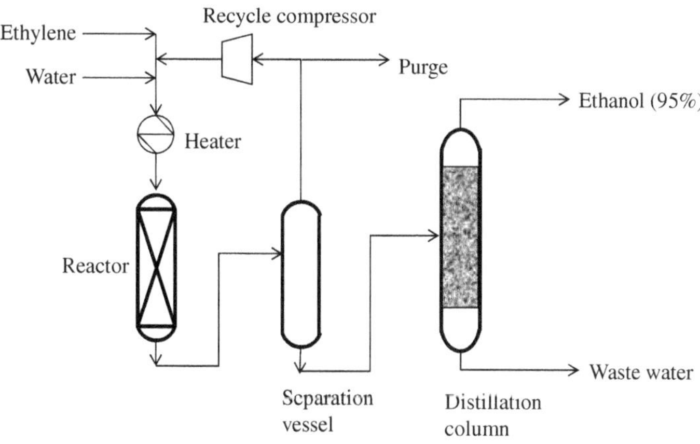

Figure 2.14. Production of Ethanol from Ethylene

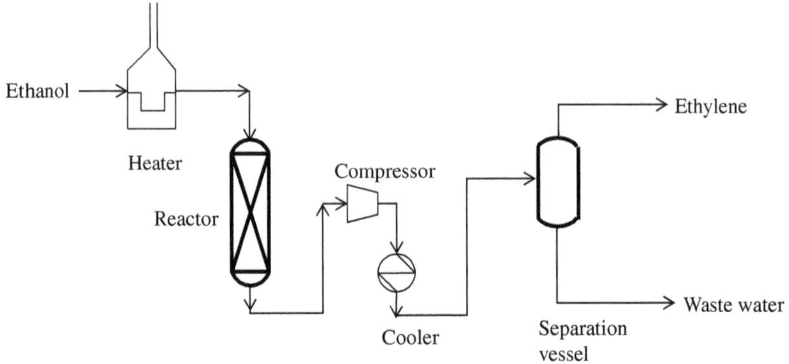

Figure 2.15.   Ethanol to Ethylene

A simplified process is illustrated in Figure 2.14. The reaction is favoured by low temperatures (below 450 K) and catalysed by acidic catalysts. Ethylene and water are heated to the required conversion temperature and passed to a reactor containing the catalyst; excess water is used. The reaction is exothermic (reaction enthalpy is −45.3 kJ/mol @ 298 K). The products are passed to a separation vessel, which separates the liquid products from unconverted ethylene, which is recycled. The liquids are passed to a distillation column, which removes hydrous ethanol (95%) from the excess water.

The reaction of ethylene with water to generate ethanol is a reversible reaction so that agricultural ethanol can be dehydrated to produce green ethylene.[35] A simplified process is illustrated in Figure 2.15. This reaction is favoured by high temperatures and is highly endothermic (reaction enthalpy +45.3 kJ/mol @ 298 K), and as a consequence, a furnace is commonly used to heat the ethanol. To maximise conversion, two or more reactors with intermediate heating are often used. Again, acidic catalysts are used. Gases from the reactor are compressed and cooled and then passed to a separation vessel, which separates the ethylene gas from the wastewater.

The production cost of ethanol or ethylene is dominated by the cost of feedstock (either ethylene or ethanol, respectively). The ideal linkage between the two is illustrated in Figure 2.16.

---

[35] *Hydrocarbon Processing* 2021 Petrochemical Processes Handbook; Technip Energies.

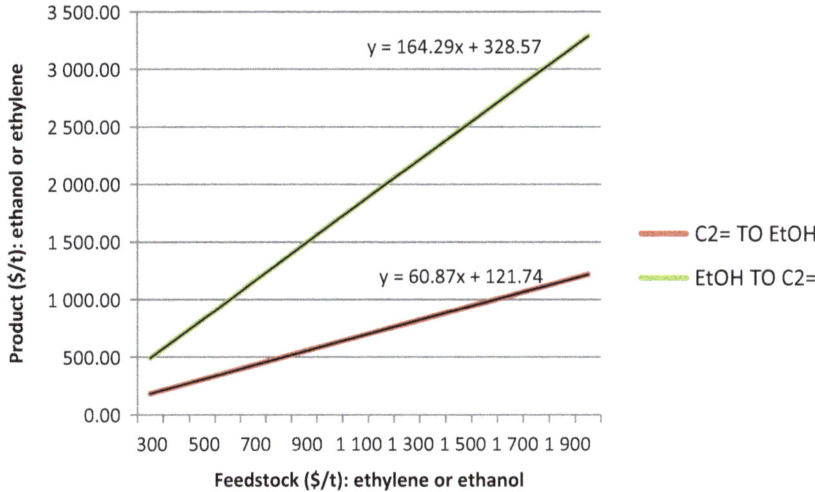

Figure 2.16. Relative Costs of Ethanol or Ethylene; Idealised Feedstock Only

The figure shows that if fermentation ethanol is available at \$800/t (say), the feedstock component (assuming ideal efficiency) of ethylene production cost will be about \$1,314/t. If this is about 80% of the total costs (allowing for capital and non-feedstock operating costs), the total production cost for green ethylene will be about \$1,642/t.

Other Routes to Ethanol

Ethylene can be produced from a range of other processes,[36] mainly as by-products. In recent years, there has been a deal of research and development in various ways to make ethanol by diverse routes. Some of the main areas of interest are as follows:

Ethanol via Synthesis Gas Fermentation

Synthesis gas is a mixture of carbon monoxide and hydrogen (CO/H_2), and this can be fermented into hydrocarbons or ethanol (alcohols) using

---

[36]A. Mohsenzadeh, A. Zamani and M.J. Taherzadah, "Bioethylene Production from Ethanol: A Review and Techno-Economical Evaluation," *Chem. Bio. Eng. Rev.* **4**(2), 75 (2017).

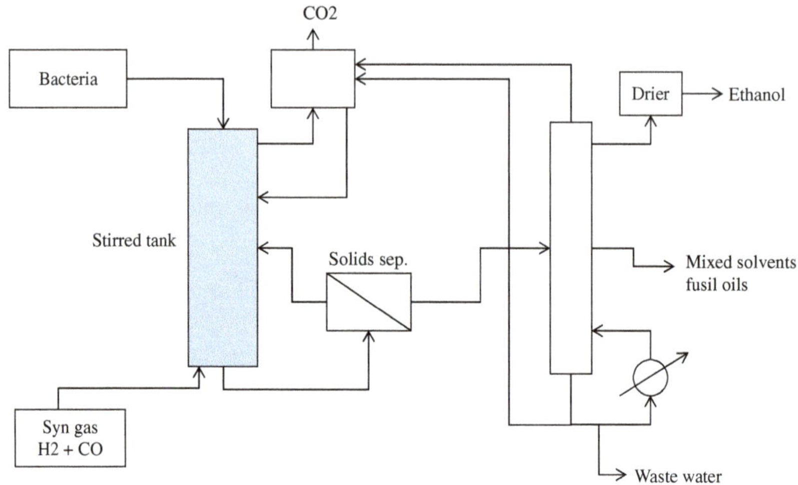

Figure 2.17.   Bio-Fermentation of Synthesis Gas

appropriate enzymes. For example, *Clostridium ljungdahlii* is shown to metabolise CO/H_2 to produce ethanol and acetic acid.

$$6CO + 3H_2O = CH_3CH_2OH + 4CO_2$$

$$2CO_2 + 6H_2 = CH_3CH_2OH + 3H_2O$$

A general outline is shown in Figure 2.17.[37]

The reaction takes place in a stirred tank reactor, which holds the bacteria/enzymes. Synthesis gas (syngas) can be produced by gasification of biomass, and this process (biomass gasification) has been the subject of much research and development[38]; it is further discussed in a later chapter. Syngas enters the bottom of the reactor. The space-time yield is claimed to be 60 g/L/h to an ethanol concentration of 36.8 g/L. The H_2 + CO

---

[37] For example, see R. Senaratne and S. Lui, Process for using biogenic carbon dioxide derived from non-fossil organic material, *US Patent*, 20130149693A1 (June 13, 2013 to Ineos Bio SA).

[38] J. Daniell, M. Kopke and S.D. Simpson, "Commercial Biomass Syngas Fermentation," *Energies* **5**, 5372–5417 (2012) quoting D.W. Griffin and M.A. Schultz, "Fuel and Chemical Products from Biomass Syngas: A Comparison of Gas Fermentation to Thermochemical Conversion Routes," *Environ. Prog. Sustain. Energy.* **31**, 219 (2012).

uptake was measured at 2 mole/min. Carbon dioxide produced in the fermentation process is discharged from the system. Solids and liquids are withdrawn from the base of the reactor and the solids are separated and passed back to the reactor, and the liquids, dominantly water, are passed to a distillation column, which separates the ethanol and by-products, such as other solvents (butanol etc.).

Fischer–Tropsch Routes

The Fischer–Tropsch process concerns the conversion of synthesis gas into hydrocarbons and water, which can be subsequently converted into fuels. This technology to produce SAF is discussed in more detail later. The process is conducted on a very large scale in South Africa, and in the version used there, the Fischer–Tropsch process produces water as a by-product. The water phase contains significant quantities of ethanol, which can be extracted.

Ethanol to Sustainable Aviation Fuel (e-Jet)

Ethanol can be easily dehydrated to ethylene (reverse of industrial ethanol production). This has been practiced in the past, but nowadays ethylene production is dominated by cracking of naphtha and light hydrocarbon gases such as ethane. Bio-derived ethylene can be converted into higher hydrocarbons boiling in the jet-fuel range using commercially proven technology.

There are several alternative approaches to convert ethylene to SAF, including:

- Variations on the Ziegler process to produce polymers but truncating the polymerisation step to produce an oligomer boiling in the jet-fuel range (typically C_8 to C_{14}). This is followed by hydrogenation to remove olefins and isomerisation to branch the carbon chains.
- The Shell Higher Olefins Process (SHOP), which uses a two-step process involving oligomerisation over a nickel catalyst followed by hydrogenation to remove olefins and isomerisation to branch the carbon chains.

- A variation on the ExxonMobil MOGD process to oligomerise light olefins to the required chain length, followed by hydrogenation to remove olefins and isomerisation to branch the carbon chains.

Ziegler Process[39]

In this process, ethylene is inserted into an alkyl-aluminium bond of triethylaluminium which is then repeated. The reaction takes place at a relatively low temperatures (90°C–120°C) at about 100 bar.

$$Al - C_2H_5 + H_2C = CH_2 = Al - CH_2 - CH_2 - C_2H_5 + nH_2C = CH_2$$
$$= Al - \{CH_2 - CH_2\}_n - CH_2 - CH_2 - C_2H_5$$

The ethylene oligomers are released by raising the temperature ($>200°C$) and lowering the pressure (50 bar). The olefins produced are straight-chain alpha-olefins with an even number of carbon atoms. Furthermore, the process can be coupled with a trans-alkylation step to increase the yield of the C_{12} to C_{18} cut sought for higher alcohols for detergent manufacture, and presumably the C_8 to C_{14} cut if SAF is to be produced. The olefins required are separated by distillation. There are several variants on the process.

For SAF, a hydrogenation and isomerisation step is required to reduce the alpha olefin to a branched paraffin suitable for jet-fuel use.

SHOP Process

The SHOP process aims to produce high-valued olefins in the C_{10} to C_{14} carbon range. These materials are sought to produce high-value detergents. The carbon range can be easily extended to cover C_8 to C_{15} hydrocarbons for optimum jet-fuel production. In outline the process for C_{10} to C_{15} range is shown in Figure 2.18.

Ethylene is added to a mixing tank where a catalyst is added along with recycled materials. This passes to the oligomerisation vessel (OLIG), where ethylene is polymerised into higher olefins, typically in the range of C_4 to C_{20} olefins. A separation unit separates unconverted ethylene and lighter and heavier olefins for recycling and the required C_{10} to C_{15}

---

[39] H.-J. Arpe, *Industrial Organic Chemistry*, 5th ed., Wiley-VCH, 2012.

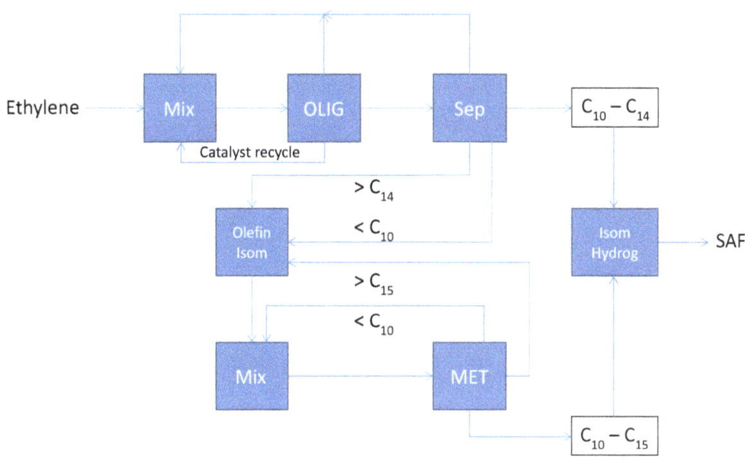

Figure 2.18. Possible Use of the SHOP Process for Producing SAF

fraction. The fraction lower than C_{10} and higher than C_{14} is passed to an olefin isomerisation unit. The duty of this unit is to isomerise the alpha-olefins produced in the ethylene oligomerisation unit into internal olefins prior to metathesis (MET). The internal olefins produced are then mixed with recycle streams and passed to the MET unit. This unit breaks the olefins at the double bond and recombines them with similar moieties from other olefins to reform an olefin with an internal double bond. The chain lengths of the olefins are scrambled.

One point to note is that oligomerisation produces olefins with an even number of carbon atoms, but after MET, there is an even distribution of chains with even and odd numbers.

For SAF, the olefins of pertinent chain size (C_8–C_{15}) are separated and other olefins are recycled as appropriate. The SAF precursors would be hydrogenated and isomerised to the SAF properties required.

MOGD Process

Another approach to the production of SAF using olefins is the Mobil Olefins to Gasoline and Distillate (MOGD) process[40]; now proposed by

---

[40] W.E. Garwood and W. Lee, *US Patent* 4,227,992 October 14, 1980 (to Mobil Oil Corporation).

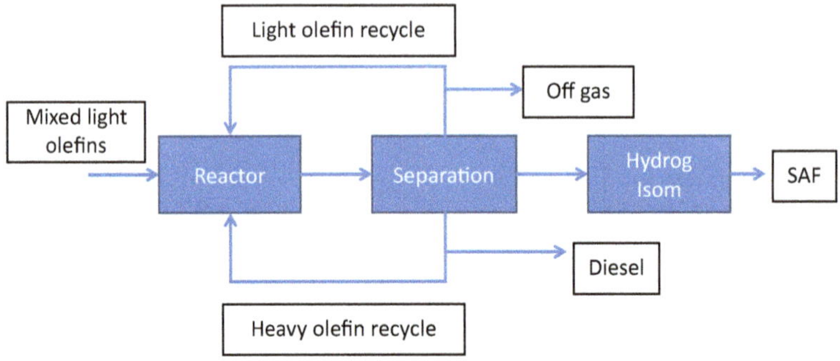

Figure 2.19. SAF by ExxonMobil Process

ExxonMobil for the production of SAF. This has been adapted to focus more on the production of the jet-fuel cut[41]; the scheme is outlined in Figure 2.19.

Mixed light olefins (C_2–C_5) produced in any sustainable manner, enter the reactor system and are oligomerised to higher olefins boiling in the gasoline to diesel range. This is passed to a separation (distillation) section. Light olefins are recycled to the reactor with a purge to remove cracked gases. The jet cut is hydroisomerised to produce the SAF. Heavy olefins are also recycled but are slow to react, so part of this stream is drawn off to produce green diesel.

The main issue with this approach is the production of the mixed olefin feedstock. Ethylene is slow to react to the preferred olefins, which are propylene and higher olefins. This mixture can be produced using the Mobil Methanol to Olefins Process with the methanol being green methanol produced by steam reforming of glycerol (a by-product of biofuel production) or steam reforming of other light hydrocarbons produced from biomass. In theory, ethanol could be used, but this generally reacts slower than methanol and has not been fully demonstrated.

Specification

These routes from ethanol via ethylene produce branched paraffins in the correct boiling range for jet fuel (C_8–C_{15} range) and sufficient branching

---

[41] J.H. Beech Jr., H. Owen, M.P. Ramage and S.A. Tabak, *US Patent* 5,073,351, Dec. 17, 1991 (to Mobil Oil Corporation).

to achieve the low pour point of jet A (−40°C) but would have too low a density for the IATA specification. Use of this fuel would limit the range of the aircraft. In other words, it would not be a drop-in fuel rather a blend stock.

The use of this type of kerosene as aviation fuel is to blend it with petroleum-produced kerosene of suitable high density and produce a final product that falls within the Jet-A or Jet-A1 specifications. Blending is well known and practiced by the fuels industry.

Sustainability

The route to SAF from ethanol places suitable arable land in competition with land for food crops (corn, beet and sugar) and land for fuel (gasoline). The concern with the amount of land needed for fuel crops has caused several jurisdictions to limit land use for fuel crops and promote the conversion of non-food biomass into ethanol. This mainly involves the conversion of lignocellulose from forestry products and the like into ethanol. This requires new fermentation media, and there has been some progress in this field, but it has been slow.

To quote Wikipedia[42]:

Although the global bioethanol market is sizable (around 110 billion liters in 2019), the vast majority is made from corn or sugarcane, not cellulose. In 2007, the cost of producing ethanol from cellulosic sources was estimated ca. USD 2.65 per gallon (€0.58 per liter), which is around 2–3 times more expensive than ethanol made from corn. However, the cellulosic ethanol market remains relatively small and reliant on government subsidies. The US government originally set cellulosic ethanol targets gradually ramping up from 1 billion liters in 2011 to 60 billion liters in 2022. However, these annual goals have almost always been waived after it became clear there was no chance of meeting them. Most of the plants to produce cellulosic ethanol were cancelled or abandoned in the early 2010s. Plants built or financed by DuPont, General Motors and BP, among many others, were closed or sold. As of 2018, only one major plant remains in the US.

---

[42] Wikipedia, Wikimedia Foundation, *Cellulosic Ethanol*, downloaded July 13, 2022.

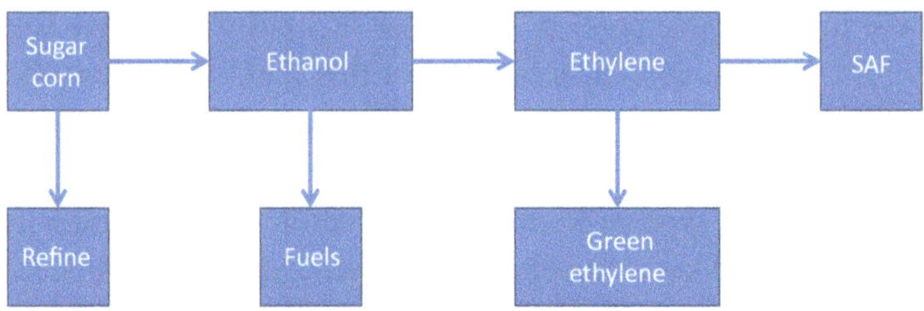

Figure 2.20. Route from Sugar/Corn to SAF

The conclusion is that at this time SAF routes requiring ethanol will be in competition with fuel ethanol and food (corn and sugar).

Economics

For this route, there are several competing products to SAF, which will influence the cost. Figure 2.20 illustrates the overall course of feedstocks from sugar or corn to SAF.

The main source of ethanol is sugar or corn. So, at this stage, the cost of the feedstock is determined by the market demand for sugar or corn syrup for the food and beverage markets. Relative to fuel, these are high-value markets. Upon conversion to ethanol, the facility owner has the option to supply the existing fuel market, that is, as a blendstock to gasoline. This market is increasing in size due to the various jurisdictions increasing the mandates for using renewable fuels in their domestic vehicle fleets. Because ethanol is used as a gasoline blendstock, the value of the ethanol is related to the prevailing price of gasoline, hence crude oil. The correlation of ethanol and gasoline prices is illustrated in Figure 2.21.

As can be seen, the correlation is positive with a reasonable correlation factor (R^2) > 0.65.

The conversion of ethanol to ethylene is a facile process. Once made, the ethylene would be classified as "green ethylene", meaning it was produced sustainably from non-fossil fuel sources. Ethylene is easily converted into commodity plastics such as HDPE, which, if produced by this route, could also be classified as "green HDPE". Although "green ethylene" would have a production cost of about 60% higher than fossil fuel-derived ethylene, the demand for green-HDPE could justify this premium.

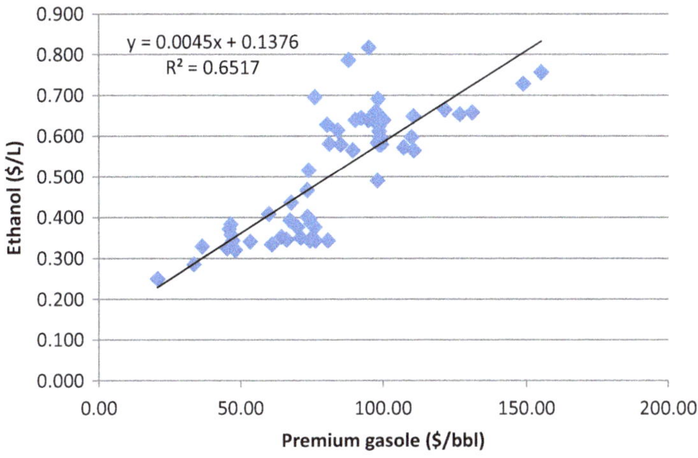

Figure 2.21. Correlation between Ethanol Price and Gasoline Price

Ethylene for SAF production would notionally have to compete against this use, raising the cost of ethylene as a feedstock for SAF production.

Case Study

Table 2.16 gives an estimate for the production cost of SAF from ethanol via ethylene. The case is based on a Ziegler polymerisation-type process producing 300 k/y SAF.[43] It is expected that the other processes will have similar outcomes.

The scaled and 2022 capital costs are estimated at $437 million for a standalone (greenfield) facility.

Non-feedstock operating costs are made up of labour charges (2% Capex/y); maintenance (3% Capex/y) and other charges (2% Capex/y). The last mentioned will be relatively high due to the higher cost of catalysts of these types of processes.

For SAF, hydrogen is required to convert the alpha-olefins produced into paraffins and isomerise the linear paraffin chains sufficiently into

---

[43] Hydrocarbon Processing, Petrochemical Processes Handbook 2005, Axens higher olefins process scale to 300 kt/y and 2022 costs for ISBL operation. Hydrogenation and isomerisation are assumed to be included in the costs.

Table 2.16. Estimate for Producing 300 kt/y SAF from Ethylene Produced via Ethanol from Sugar/Corn Fermentation

Capacity (product)	kt/y	300.00
SAF density (as C_{12})	kg/L	0.750
SAF	t/y	300.00
	ML/y	400.00
	bbl/d	6897.16
CAPEX 2022	$M	437.55
Return on capital (2, 10, 20)[a]	% Capex/y	13.84%
Return on capital	$M/y	60.56
CAPEX 2022	$/bbl	24.05
Labour (2% capex)	$M/y	4.38
Maintenance (3% capex)	$M/y	13.13
Other (2% capex)	$M/y	8.75
OPEX	$M/y	26.25
	$/bbl	10.43
Hydrogen	kt/y	4.70
	$/t	3,000
	$M/y	14.11
	$/bbl	5.61
Ethylene	t/y	345.00
Ethylene price	$/t	777.88
	$M/y	283.25
	$/bbl	112.51
Diesel by-product	t/y	36.00
Price ($110/bbl)	$/t	817.3
Credits	$M/y	29.42
	$/bbl	11.69
Net production cost	$M/y	339.86
	$/bbl	135.00

[a]ROC for 2-year construction, 10% Dcf, 20-year plant life

product with suitable cold flow properties. The hydrogen requirement is estimated by the quantum of product produced (SAF and by-products) with a price of $3,000/t, which is typical for refinery operations. Green hydrogen may cost significantly more than this.[44]

The process operates at about 85% efficiency on ethylene, so it requires 345 kt/y ethylene. The ethylene cost is taken at $778/t, which is aligned with a typical ethanol trading price of $0.614/L and crude oil at $100/bbl (as illustrated in Figure 2.16). It is assumed that the venture to make SAF will not be tempted to sell the ethylene on the market, which has a much higher value.

The process produces various by-products, which are assumed to be reduced and isomerised into diesel. This generates a by-product credit, which serves to reduce the production cost to an estimate for SAF of $135/bbl.

The breakdown of costs is illustrated in Figure 2.22. The figure clearly shows the dominant cost is for ethylene feedstock (71% of the costs), which in turn means the cost of production of ethanol.

The sensitivity of the estimate in SAF costs is shown in Figure 2.23. As noted above, the opportunity cost to produce ethylene is much higher

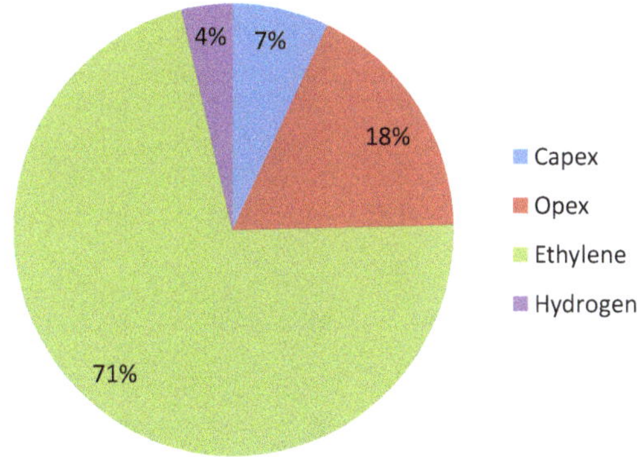

Figure 2.22. Breakdown of SAF Costs for the Case Study

---

[44] D. Seddon, *The Hydrogen Economy — Fundamentals, Technology, Economics*, World Scientific, 2022.

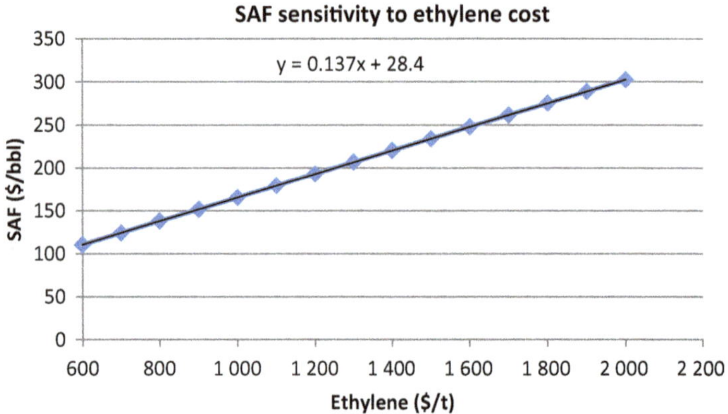

Figure 2.23. Sensitivity of SAF Production Costs to Ethylene Cost

than the value used in this case study. Using an opportunity value for the ethylene of $1300/t would increase the SAF cost to $206/bbl.

With crude oil at the $100/bbl mark and commercial jet fuel at about $110/bbl so that even with an optimistic approach to the ethanol and ethylene cost, SAF production is significantly higher than conventional jet fuel.

3

SAF FROM VEGETABLE OILS AND FATS

Production of Vegetable Oil

Vegetable oils and fats are extracts from seeds and nuts. These comprise a complex mixture of chemicals. Crushing the seed or nut expels the oil. For biofuel applications, further oil is extracted from the crushed seed by solvent extraction using an organic solvent such as hexane. Currently, the prime use of the oil is to produce biodiesel and green diesel rather than gasoline or jet fuel.

The oil comprises substantially fatty acid esters of glycerol into which are dissolved minor components of oil-soluble materials such as phospholipids. The small amounts of non-glycerides present may ultimately cause contamination of the final biofuel with trace elements such as phosphorus and sulphur.

In principle, any vegetable oil could be converted into the feedstock or a blending component for biofuel. However, certain properties of vegetable oils could detract from the final oil quality for biofuel production, which will then in turn contaminate the final biofuel and degrade the fuel quality. This restricts the selection of vegetable oils suitable for biofuels, nevertheless, there is a wide range of feedstocks that can be used. Of the range of feedstocks available, rapeseed[1] (and members of the rape

---

[1] Rapeseed is widely used in Europe. Rape itself contains a high level of erucic acid and is grown for this purpose. The low erucic acid version is generally called Canola outside Europe. There is some ambiguity in European data as to which form is being referenced.

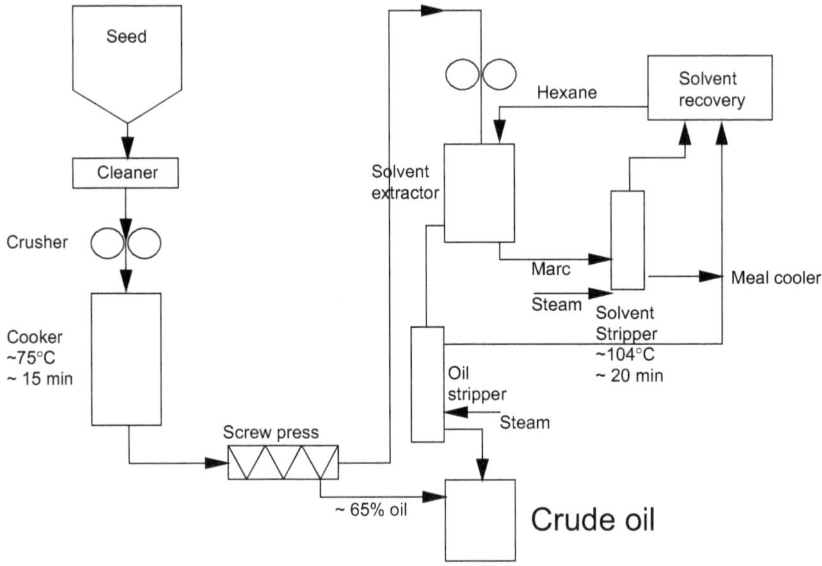

Figure 3.1. Typical Production of a Crude Vegetable Oil

family, particularly Canola), soy and palm oil are significant feedstocks for biofuel.

In addition to vegetable oils, animal fats can be used. This is available in large quantities in countries with large meat processing industries, for example, tallow from lamb fat. The production of Canola oil is typical and is illustrated in Figure 3.1.

Seed enters the crushing plant via a hopper. A screen cleaner filters out extraneous material. The seeds are then crushed, and the crushed seed is passed to a cooker. This heats the seed to 75°C for typically 15 minutes and helps expel oil from the seed structure. The hot seed is now passed to a screw press, which expels the bulk of the oil.

The residue is then passed to the solvent extractor. First, it is further crushed before entering an extraction vessel where it is contacted with hexane. Oil dissolves in the hexane, and the liquid mixture is passed to a solvent stripper, which recycles the solvent and collects the second portion of crude oil.

The solid residue from the solvent extractor (known as marc) is heated with steam to expel entrained solvent. The final solids (known as meal) are passed to a cooler.

This produces crude vegetable oil. This contains a range of oil-soluble materials, which would cause problems in producing specification biofuel. To remove these, the crude oil is subject to a series of washing operations in order to produce the clean oil, which is suitable for biofuel production. For Canola (rapeseed), typical washing operations are

- Gums are removed by (i) hot water washing, (ii) acid treatment, which precipitates the gum and (iii) centrifuging to separate the gum from the oil.
- Free fatty acids are neutralised with caustic soda. This produces soap, which is washed out of the oil with hot water.
- Water from the wash steps is removed from the oil by centrifuges and drying sieves.
- Coloured matter (carotenoids) is removed using an earth filter.
- For some oils, odours are removed by vacuum steam distillation. This is not used for some types of oil (rapeseed).
- For food use, permitted antioxidants are added.

After these processes, the vegetable oils comprise substantially fatty acid triglycerides with only about 2% or less non-triglyceride material. The non-triglycerides comprise mainly sterols, alcohols (tocopherols) and high molecular weight hydrocarbons (e.g. squalene). This is often referred to as unsaponifiable matter. After washing operations, the vegetable oil is ready for the production of biofuel and, in particular, biodiesel.

Structure of Vegetable Oils and Fats

For biofuel, the main interest is the structure of the fatty acid portion of the triglyceride. There are three fatty acid moieties attached to the glycerol moiety. These fatty acid moieties are generally

- linear chains with few, if any branches,
- have even carbon numbers,
- occur with a range of carbon numbers which is dependent on the source feed,

- contain few unsaturated links,
- the location and growing conditions influence the final composition.

Thus, the detailed composition of the fatty acids is different for each individual feed type — soybean, palm oil and rapeseed. There are minor changes with the growing season and location. High levels of unsaturation (high olefin content measured by iodine number [IN]) are less preferred than saturated chains, which have better stability in storage. However, saturated chains tend to deliver products with higher fuel pour points and poor cold climate properties.

Fatty Acid Methyl Esters — Biodiesel

Although there is an extensive community of supporters for the use of vegetable oils as diesel fuel or diesel blends, the use of vegetable oils or fats is not approved by original engine manufacturers (OEMs). The main concern is the control of impurities, and the high viscosity of the oils leads to erosion and corrosion of engine components and poor combustion. OEMs recommend the conversion of vegetable oils and fats into fatty acid methyl esters (FAME) in which impurities are reduced and the viscosity is more amenable to components such as high-pressure pumps and injectors.

There are various technologies to produce suitable biodiesel. In essence, these various technologies involve contacting the vegetable oil with methanol and a catalyst, which brings about transesterification of the triglycerides to form FAME.

There is no change in the structure of the fatty acid chains by this process.

The crude FAME is contaminated with glycerol, water, free acids and catalyst. Potentially the 2% or less impurity in the oil feedstock could contaminate the final biodiesel, especially regarding the introduction of elements other than carbon, hydrogen and oxygen. These are removed by extensive washing procedures. These contaminants can cause corrosion and other problems in the engine, and limits are placed on the levels in the final biodiesel.

The control of elemental impurities (sulphur, phosphorus, alkali and alkaline earth metals) is required to minimise erosion and deposit formation on engine parts. Effective control of minor impurities is, for the most part, a function of the operation of the biodiesel production plant.

A discussion of the potential of biodiesel contamination by the small quantum of impurities in the triglycerides will require a detailed analysis of the source feedstock and knowledge of the specific production plant employed.

FAME produced in this manner is generally referred to as biodiesel.

Principal Feedstocks — Soybean, Rapeseed, Palm Oil and Tallow

In theory, any vegetable or fat oil can be used to produce biodiesel. Hence, there is an enormous list of potential source oils. For widespread use of a particular vegetable source oil, the production should minimise the cost of the feedstock by maximising the crop yield and oil yield per hectare. Pertinent yields are given in Table 3.1.

Table 3.1. Possible Biodiesel Crops[2]

Crop	kg oil/ha
Algae (actual yield)	6,894
Algae (theoretical yield)	39,916
Avocado	2,217
Brazil nuts	2,010
Calendula	256
Camelina	490
Cashew nut	148
Castor beans	1,188
Chinese tallow	5,500

(Continued)

---

[2] Adapted from Wikipedia, Wikimedia Foundation, *Table of Biodiesel Crop Yields*, downloaded September 20, 2023.

Table 3.1. (*Continued*)

Crop	kg oil/ha
Cocoa (cacao)	863
Coconut	2,260
Coffee	386
Copaiba	10,000
Coriander	450
Corn (maize)	145
Cotton	273
Euphorbia	440
Hazelnuts	405
Hemp	305
Jatropha	1,590
Jojoba	1,528
Kenaf	230
Linseed (flax)	402
Lupine	195
Macadamia nuts	1,887
Mustard seed	481
Oats	183
Olives	1,019
Opium poppy	978
Palm oil	5,000
Peanuts	890
Pecan nuts	1,505
Pumpkin seed	449
Rapeseed (Canola)	1,000
Rice	696
Safflower	655
Sesame	585
Soybean	375
Sunflowers	800
Tung oil tree	790

The list can be culled due to

- some oils have very high innate value for other uses, for example, olive oil.
- some oils are well known to be unsuitable for biodiesel. For example, castor oil, which has a very high viscosity, and linseed oil, which has a very high level of unsaturated material. These two oils have other uses — lubrication and lacquers, respectively.
- some vegetable oils are produced in volumes too low for their cultivation for biodiesel.

The oils listed below are primarily used for other purposes (all but tung oil are edible) and all have been considered for use as biofuel. Oils of particular interest are in bold in Table 3.1.

- Castor oil is lower cost than many candidates. However, the oil has an unusual structure with the fatty acid chains containing a hydroxyl moiety. This causes a significant increase in viscosity, which leads to its use as a lubricant rather than its use in biofuels production.
- **Coconut oil** (copra oil) is promising for local use in places that produce coconuts. Coconut oil is high in C_{14} fatty acids, which makes it an ideal crop for the production of biojet fuel (sustainable aviation fuel [SAF]) rather than motor diesel.
- Corn oil is appealing because of the abundance of maize as a crop.
- Cottonseed oil is produced in large volumes as a by-product of cotton production, and the seed could be available for biodiesel production.
- Linseed oil (flax), from *Camelina sativa*, was used in Europe in oil lamps until the 18th century. The oil is highly unsaturated, which leads to its use in producing lacquers and paints.
- Hemp oil is relatively low in emissions and has a high flash point. Production is problematic in some countries because of its association with marijuana.
- Mustard oil is shown to be comparable to Canola oil as a biofuel.
- **Palm oil** is very popular for biofuel due to high crop yield, but the environmental impact of growing large quantities of oil palms has recently called the use of palm oil into question.

- Peanut oil was used in one of the first demonstrations of the diesel engine in 1900.
- Radish oil, wild radish, contains up to 48% oil, making it appealing as a fuel.
- **Rapeseed oil** is the most common base oil used in Europe in biodiesel production. The main variety is Canola (low erucic acid rapeseed). Recently there has been an increasing interest in rapeseed varietals containing large quantities of erucic acid (e.g. carinata), which potentially can give higher yields of SAF.
- Ramtil oil is used for lighting in India.
- Rice bran oil is appealing because of its lower cost than many other vegetable oils. Widely grown in Asia.
- **Safflower oil** has been explored as a biofuel in Montana.
- **Soybean oil** is not directly economical as a fuel crop but is widely used because it is a by-product of soybean production for food uses.
- **Sunflower oil** is suitable as a fuel but not necessarily cost-effective.
- **Tung oil** is referenced in several lists of vegetable oils that are suitable for biodiesel.

Inedible Oils Used Only or Primarily as Biofuel

These oils are extracted from plants that are cultivated solely for producing oil-based biofuel. These, plus the major oils described above, have received much more attention as fuel oils than other plant oils.

Algae oil[3] has been of interest in the production of fuels since the early 1940s,[4] when it was realised that algae cultivation could give high yields of oil. Current developments include carbon dioxide capture to increase yields. So far, the promise of low-cost production of oil using algae has failed to materialise.

---

[3] P. Schlagermann, G. Gottlicher, R. Dillschneider, R. Rosello-Sastre and C. Posten, "Composition of Algae Oil and its Potential as Biofuel," *J. Combustion*, 2012, Article ID 285185. Publication date - 09 April 2012.

[4] R. Harder and H. von Witsch, The Growth Factors involved in Microalgae Cultivation for Biofuel Production: A Review, *Forschungsdienst Sonderheft.* **16**, 270–275 (1942) *ibid. Berichte der Deutschen Botanischen Gesellschaft.* **60**, 146–152 (1942).

Copaiba[5] is an oleoresin tapped from species of the genus *Copaifera*. Used in Brazil as a major source of biodiesel. It is of interest because of its claimed high oil yield.

Honge oil[6] (also known as Pongamia or Karanja) was pioneered as a biofuel in Bangalore, India.

Jatropha oil[7] is widely used in India as a fuel oil. Has attracted strong proponents for use as a biofuel. Regarded as a noxious weed in many jurisdictions.

Milk bush[8] was popularised by chemist Melvin Calvin[9] in the 1950s and researched in the 1980s by Petrobras, the Brazilian national petroleum company.

Technology to Produce FAME

As noted above, the direct use of vegetable oils and fats is not supported by OEMs. The principal problems stem from the high viscosity of the oils, which cause increased erosion problems with pumps and injectors and poor spray formation in the cylinder, resulting in less efficient combustion. To solve this problem, the vegetable oils are transesterified with methanol to convert the glycerol esters to FAME, which have

---

[5] Wikipedia, *Copaifera langsdorffii*, downloaded September 20, 2023.

[6] N.R. Banapurmath, P.G. Tewari and R.S. Hosmath, "Effect of Biodiesel Derived from Honge Oil and its Blends with Diesel When Directly Injected at Different Injection Pressures and Injection Timings in Single-Cylinder Water-Cooled Compression Ignition Engine," *Proc. Inst. Mech. Eng. Part A: J. Power Energy* **223**(1), 31–40 (2009). doi:10.1243/09576509JPE673.

[7] T.M.I. Riayatsyah et al., "Current Progress of Jatropha Curcas Commoditisation as Biodiesel Feedstock: A Comprehensive Review," *Front. Energy Res.* **9**, (2022). doi:10.3389/fenrg.2021.815416.

[8] O. Ogunkunle, O.O. Oniya and A. O. Adebayo, "Yield Response of Biodiesel Production from Heterogeneous and Homogeneous Catalysis of Milk Bush Seed (*Thevetia peruviana*)," *Energy and Policy Res.* **4**, 21–28 (2017).

[9] M. Calvin, *Fermentation and Hydrocarbons*, II Latin-American Botanical Congress, January 22–27, 1978.

Table 3.2. Comparison of Mineral Diesel with Typical FAME and Vegetable Oils

Property	Mineral diesel	Jatropha FAME	Jatropha oil
Density (kg/m^3)	840	879	917
Kinematic viscosity at 40°C (cSt)	2.44	4.84	35.98
Pour point (°C)	6	3	4
Flash point (°C)	71	191	229
Conradson carbon residue (%w/w)	0.1	0.01	0.8
Ash Content (%w/w)	0.01	0.13	0.03
Calorific value (MJ/kg)	45.34	38.5	39.07
Sulphur (%w/w)	0.25	<0.001	0.0
Cetane number	48–56	51–52	23–41
Carbon (%w/w)	86.83	77.1	76.11
Hydrogen (%w/w)	12.72	11.81	10.52
Oxygen (%w/w)	1.19	10.97	11.06

lower and acceptable viscosity, though still higher than conventional mineral diesel.

Table 3.2 gives a comparison of the pertinent properties of mineral diesel, FAME derived from jatropha, and jatropha oil,[10] illustrating the above points.

The process is illustrated figuratively in Figure 3.2, which illustrates the conversion of a vegetable oil molecule into FAMEs and glycerol.

The FAME mixture replicates the fatty acid distribution and the olefin content and distribution of the original feedstock. The FAME methyl ester mixture is commonly referred to as biodiesel. In a pure (as made) form, this is commonly called B100 (100% biodiesel). If the FAME is mixed with mineral diesel, say a 20% FAME/80% mineral diesel mix, this is referred to as B20. Common mixes are B5 (5% FAME), B7.5 (7.5% FAME) and B10 (10% FAME).

Since the hydrocarbon chain is important in determining the character of the FAME, the nomenclature adopted to describe the chain follows the practice used in the oleochemicals industry in general. This describes the

---

[10] D. Luna et al., "Technological Challenges for the Production of Biodiesel in Arid Lands," *J. Arid Environments* **102**, 127–138 (2014).

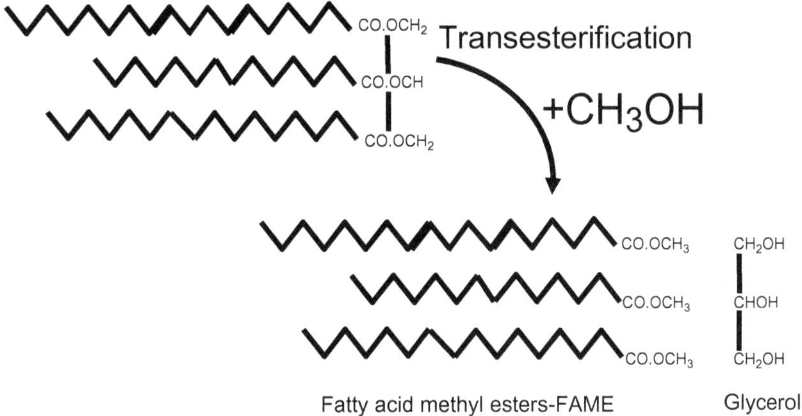

Figure 3.2. Illustration of Trans-Esterification to Produce FAME and Liberate Glycerol

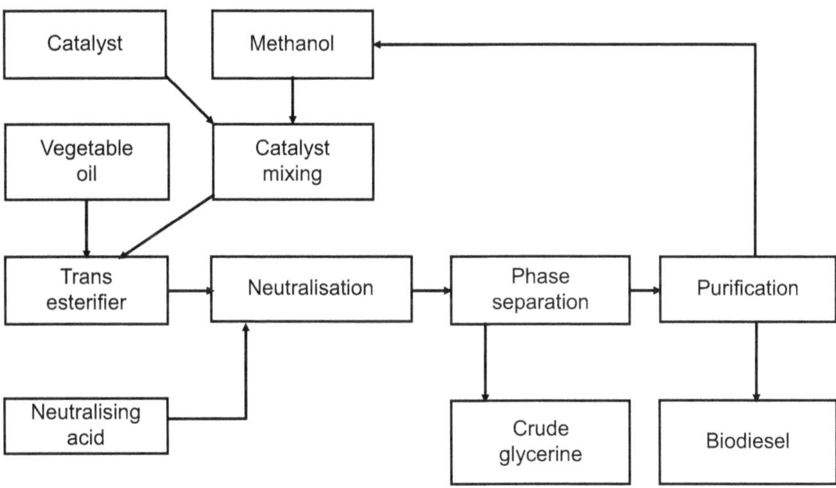

Figure 3.3. General Approach to Transesterification of Vegetable Oils to for Biodiesel

chain length — carbon number, for example, C18 — followed by a second number indicating the number of olefinic bonds — degree of unsaturation, for example, C18:1.

Biodiesel and mixes with mineral diesel are not the same as "green" diesel, which is discussed in more detail later in this work. There are several technologies to produce biodiesel. The general approach is illustrated in Figure 3.3.

The vegetable oil is charged to a transesterification vessel where it is mixed with methanol and a catalyst (this can be acidic or alkaline). The mixture is stirred, which brings about the reaction. Excess acid or alkali is neutralised (acid if an alkali catalyst is used). The mixture is passed to a phase separator where the crude glycerine (also called glycerol) is separated, and then the biodiesel is passed to a final purification stage.

The key issues to produce the FAME process are

- removal of water, as this promotes engine corrosion.
- removal of free fatty acids, as this produces soap, which interferes with the separation and promotes engine corrosion.
- removal of insoluble compounds (phospholipids etc.) as these cause filter blockages and promote instability in the fuel.

Note that FAME is not distilled. The boiling point of the methyl esters is higher than the diesel range, and many of the methyl esters decompose on distillation. Olefins in the FAME decompose easily at high temperatures, which results in fouling of equipment.

Glycerin or Glycerine or Glycerol[11]

The main use of glycerol is in pharmaceuticals and cosmetics. Biodiesel production produces a crude product, and further refining is necessary. More value is attained using pure vegetable oils not contaminated with animal products (so-called kosher glycerin).

The large increase in biodiesel production in recent years has served to depress glycerol prices, and there is active research and development in alternative uses for the glycerol by-product. One such development is the potential for selective hydrogenation to produce propylene glycol.[12]

---

[11] The terms glycerin, glycerine and glycerol are used interchangeably in this work. Different parts of the downstream industry using this by-product may prefer one or the other of the terms.

[12] *Hydrocarbon Processing*, 2021 Petrochemical Processes Handbook; Johnson Matthey, Propylene Glycol Process.

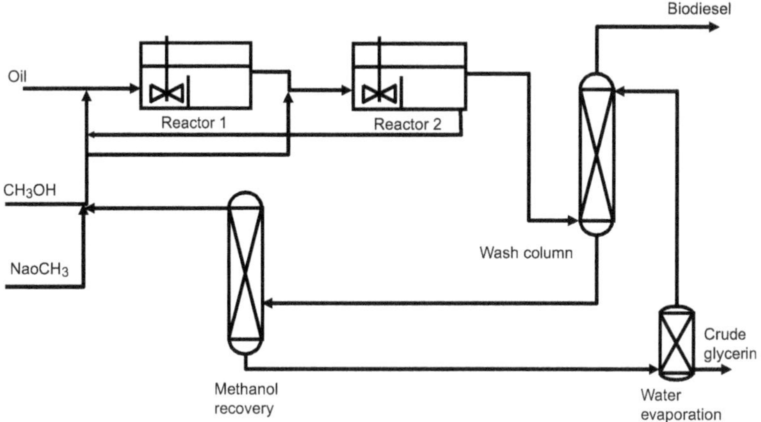

Figure 3.4.　Typical Stages in FAME (Biodiesel) Manufacture

Figure 3.4 illustrates the layout of a typical[13] facility.

A vegetable oil is passed to a stirred tank reactor, to which methanol and sodium methanate, which acting as the transesterification catalyst, are also added. The product is passed to a second reactor where more methanol and catalyst are added to complete the conversion of the oil. The product from the second reactor is passed to a wash column, which separates the lighter biodiesel from the water phase containing unconverted methanol and glycerin. The water phase is passed to a distillation column, which removes the unconverted methanol for recycling. The water phase is passed to an evaporator with the duty to strip the crude glycerin of water, which is recycled to the wash column.

Biodiesel and Biodiesel Standards

Various jurisdictions have introduced standards for biodiesel. The Australian standard is typical, and the controlled parameters are given in Table 3.3.

The nature of the feedstock will influence the parameters shown in Table 3.4. Ignoring the minor impurities, this mainly concerns the nature of the fatty acids involved.

---

[13] Based on the Lurgi transesterification process.

Table 3.3. Fuel Standard (Biodiesel) Australian Determination 2003

Property	ASTM or EN test	AUS B100
Sulphur (mg/kg)	D5453	10
Sulphated ash (ppm)	D874	200
Carbon, 10% residue (% max)	D4530	0.3
Water and sediment (ppm)	D2709	500
Phosphorus (ppm)	D4951	10
Free glycerol (% mass max)	D6584	0.02
Total glycerol (% mass max)	D6584	0.25
Alkali metals (ppm max)	EN14108/9	5
Alkaline metals (ppm max)	EN14538	5
Alcohol content (% mass max)	EN14110	0.2
Density kg/cm (range)	D1298	860–890
Distillation T90 (°C max)	D1160	360
Viscosity (cSt @ 40°C)	D445	3.5–5.0
Flash point (°C) minimum	D93	120.0
Copper corrosion max	D130	3
Ester content (% min)	prEN14103	96.5
Acid value (mgKOH/kg max)	D664	0.8
Total contamination (ppm)	D5452	24
Cetane number, min	D613	51
Oxidation stability (h @ 110°C)	EN14112	6

Table 3.4. Properties of Biodiesel Potentially Influenced by Feedstock Source

Property	ASTM or EN test	AUS B100
Carbon, 10% residue (% max)	D4530	0.3
Density kg/m^3 (range)	D1298	860–890
Distillation T90 (°C max)	D1160	360
Viscosity (cSt @ 40°C)	D445	3.5–5.0
Flash point (°C) minimum	D93	120.0
Cetane number, min	D613	51
Oxidation stability (h @ 110°C)	EN14112	6

Carbon Residue

It is not possible to predict with any degree of accuracy the likely level of carbon residue in any given FAME mixture. The amount of carbon residue may be influenced by the incorporation of permitted additives such as antioxidants in any biodiesel produced.

However, some general observations can be made. The permitted carbon residue value is likely to be exceeded for those fatty acids that are easily thermally decomposed. This primarily concerns fatty acids, which are

- large in terms of molecular weight and therefore have a corresponding high boiling point. As the sample is heated to achieve the 10% residue level, the sample reaches a high temperature, which decomposes the fatty acid.
- have a high degree of unsaturation (high level of olefins), which imparts lower thermal stability on the fatty acid.
- contain thermally unstable groups such as cyclopropenoid groups, which are found in some vegetable oils.

Density

In the finished biodiesel, the density[14] should be in the range 860 to 890 kg/m^3. The lowest FAME is likely to be methyl octanoate, which has a density of 0.876 kg/m^3. The lower density criterion is unlikely to cause an issue with exotic feedstocks.

Since FAME, density will blend in linear proportion to the mass fraction present. Using this, the final densities given later in Table 3.6 were estimated from the densities of the pure methyl esters. The densities for some of the larger FAMEs were not available (many are solids at room temperature). To take account of this, the densities were adjusted accordingly, and an estimated density was derived.

Most of the vegetable FAME is predicted to have a density below 890 kg/m^3. Tallow (lamb fat) falls outside the range with estimated values near

---

[14] Note that many biodiesel standards call for a maximum density of 900 kg/m^3.

to 900 kg/m^3. However, the literature[15] indicates that tallow ester is solid at ambient temperatures with a density of 872 g/m^3.

Since tallow has poor cold flow properties, it is generally used as a blend with vegetable FAME, which would reduce the biodiesel density.

Distillation

The maximum value for the T90 point is typically 360°C. The T(90) point is the temperature at which 90% of the biodiesel will distil at atmospheric pressure.

If a good-quality FAME analysis is available for a particular biodiesel, then the boiling character and distillation profile of a biodiesel can be simulated.[16] However, it should be noted that the distillation profile is narrow when compared to conventional petroleum diesel; most FAME boil over a narrow temperature range.

In order to get an approximation for the T(90) value and to be able to come to some view if a particular biodiesel will have an issue with the T(90) point, in Table 3.6, cognisance is taken of the normal boiling points of the main components.

The fatty acids C$_{20}$ and higher have normal boiling points above 360°C.[17] Biodiesel with quantities of these and large fatty acids is likely to compromise the T(90) distillation value. There is considerable error in the practical determination of the T(90) point, and there should be some flexibility in the regulation of this criterion.[18] The mass percent of the biodiesel boiling below 360°C has been estimated. Comparing this with the required 90% value indicates that the only oils that will compromise the T(90) level are those oils with high levels of C22:1 (erucic acid).

---

[15] B. Judd, *Biodiesel from Tallow*, for Energy Efficiency and Conservation Authority, November 2002.

[16] W. Yuan, A.C. Hansen and Q. Zhang, "Vapor Pressure and Normal Boiling Point Predictions for Pure Methyl Esters and Biodiesel Fuels," *Fuel* **84**, 943–950 (2005).

[17] M. Worgetter, H. Prankl, J. Rathbauer and D. Bacovsky, *Local and Innovative Biodiesel*, HBLFA, March 2006; these authors report lower temperatures for the normal boiling points of the FAME of interest than other sources.

[18] D. Seddon, Report to DEH "Further Analysis of ASTM Method D86 ....," March 2003.

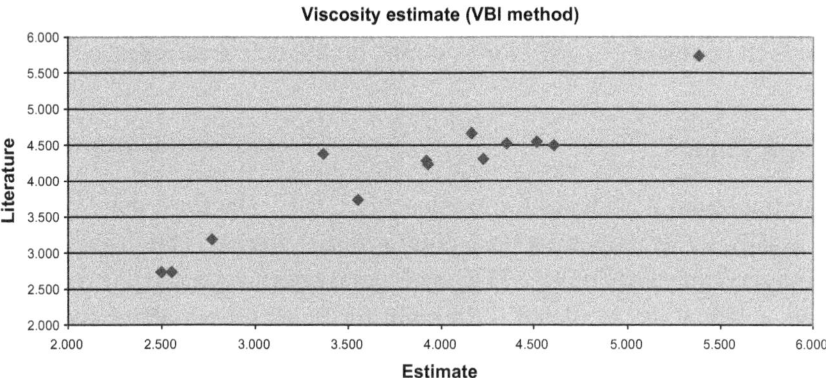

Figure 3.5. Comparison of Literature and VBI Estimates for FAME

This includes high-erucic acid rapeseed oil, which is one of the sources of biodiesel in Europe.

Viscosity

Viscosity should be in the range of 3.5 to 5.0 cSt. Viscosity is difficult to quantify, especially in mixtures. A relatively large amount of lower molecular weight fatty acid may cause an issue with the lower limit. There is a paucity of published data on viscosity.

Comparing the method of Clements[19] with available literature values[20] gives a poor result. In Table 3.6, the standard refinery viscosity blending index (VBI) method[21] was used and gave better agreement (Figure 3.5). However, it should be noted that there is a large degree of error in these estimates.

The result indicates that only those oils with a high level of lower molecular weight esters will have an issue with the lower bound viscosity

---

[19] L.D. Clements, "Blending Rules for Formulating Biodiesel Fuel," 1996; available on National Biodiesel Board website.

[20] M. Worgetter, H. Prankl, J. Rathbauer, D. Bacovsky, "Local and Innovative Biodiesel", HBLFA, March 2006.

[21] R.E. Maples, *Petroleum Refinery Process Economics*, PennWell, 1993.

of 3.5 cSt., for example, coconut, and those high levels of erucic acid at the high bound of 5.0 cSt., for example, high erucic acid rapeseed oil.

Flash Point

The minimum flash point for biodiesel is 120°C. The flash point of fuels is usually set by the lowest flash point material present. The flash point of methyl octanoate (C8:0) is 73°C and methyl decanoate (C10:0) is 93°C. This is below the regulated flash point for biodiesel but above the regulated standard for automotive diesel (typically 61.5°C). Fuels with high levels of the lighter FAME may cause an issue with the regulated flash point for biodiesel. The flash point for coconut FAME is reported to be about 100°C.

Oxidation Stability

Oxidation stability is improved by a higher degree of saturation. FAME with a high level of C18:3 and similar highly unsaturated fatty acid esters will have poor oxidative stability. It might be expected that oxidation stability would correlate with the iodine number IN. As the data in Figure 3.6 shows, this correlation is poor with many oil compositions showing poor oxidation stability. From this, it is evident that all but the most

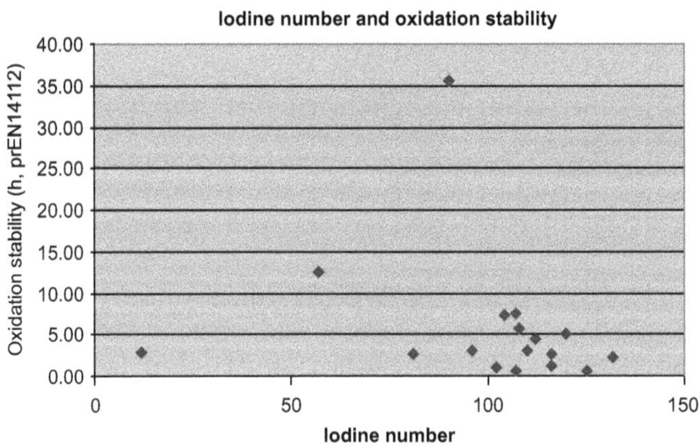

Figure 3.6.   Correlation between Iodine Number and Oxidation Stability for FAME

stable FAME (e.g. produced from tallow) would require the addition of antioxidants to achieve the desired level of oxidative stability of 6 h with the prEN14112 test method.

Cetane Number

The biodiesel standard calls for a minimum cetane of 51. Cetane is difficult to measure because of a scarcity of the required test engines, which are expensive (ASTM method D613). Biodiesel cetane values quoted in the literature may often be estimates or determined by non-approved methods. For the most part, petroleum diesel is regulated by the cetane index, which cannot be used for FAME. Methods have been developed for estimating cetane number (CN) from the saponification number (SN) and IN of the individual FAME making up the biodiesel.[22] The saponification and iodine values can be estimated from the FAME distribution.

$$SN = \Sigma(560 \times Ai)/MWi$$

$$IN = \Sigma(254 \times D \times Ai)/MWi$$

where Ai is the percentage of the ester present, MWi is the molecular weight of the ester and D is the number of double bonds in the ester.

The FAME distribution for the key fatty acids, together with an estimated CN is given in Table 3.6.

$$CN = 43.6 + 5458/SN - 0.225*IN$$

The CNs are calculated from the full fatty acid distribution. This in turn is based on a calculated SN and IN. An iodine value for many of the oils is available in the literature.

The correlation between the estimated iodine values and the literature values is shown in Figure 3.7. This illustrates the degree of error likely in the cetane estimates. As the iodine value rises above 120, there is a drift downwards in the estimate for the iodine value relative to the literature value. This is explained as due to lower levels of iodine saturation of highly unsaturated fatty acids.

---

[22] M.M. Azam, A. Waris and N.M. Nahar, "Prospects and Potential of Fatty Acid Methyl Esters of Some Non-Traditional Seed Oils for use as Biodiesel in India," *Biomass Bioenergy* **29**, 293 (2005).

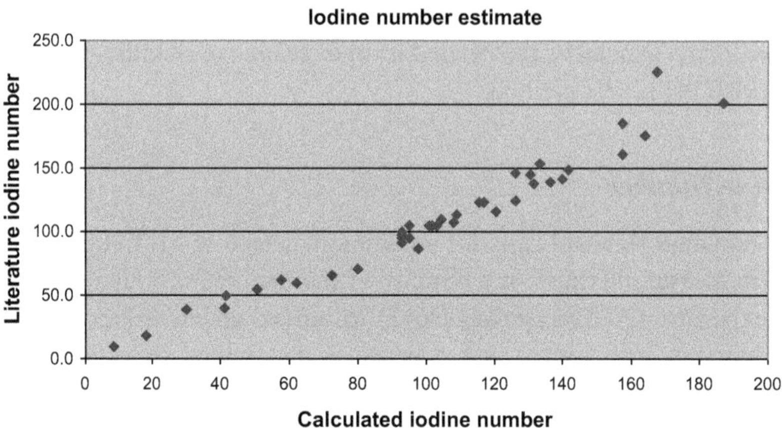

Figure 3.7.   Correlation between Literature and Calculated Iodine Numbers for FAME

Soybean and canola have estimated cetane below the regulated value of 51. Literature values for soybean methyl ester cetane are in the range of 45 to a high of 67, and rapeseed methyl ester ranges from 48 to 62.[23] Since we know that these materials produce good-quality biodiesel, in interpreting this table of estimated CNs, values higher than 43 (soybean) should be considered as adequate.

Oils in Table 3.6 with cetane values below 43 tend to have a high level of unsaturated fatty acids. These would lead to poor biodiesel on the grounds of oxidation stability and carbon residue values as well as poor cetane.

Pertinent properties for some of the principal feedstocks are shown in Table 3.5.

Table 3.6 brings together some of the fatty acid data from a diverse range of literature sources. Only the principal fatty acids are included in the table. The totals in the final column are for the fatty acids in the table; the difference between 100% can be taken to represent fatty acids that have not been included. For instance, coconut has a significant amount of fatty acid material below C16:0, namely C14:0, C12:0 and C10:0. Kapok has a significant amount of C18 cyclopropenoid fatty acids.

---

[23] J. Van Gerpen, "Cetane Number Testing of Biodiesel," Liquid Fuels and Industrial Products from Renewable Resources—Proceedings of the Third Liquid Fuels Conference, Nashville, 15-17 September 1996, pp. 166–176.

Table 3.5. Properties of Biodiesel from Different Feedstocks

		Palm[24]	Tallow	Soybean	UCO[a]	Coconut	Peanut	Canola (rape)[25]	Diesel
Density	kg/m^3	857	874	894	855–871	875	883	842	839
Viscosity	mm^2/s	4.45	4.824	4.093	4.57–5.28	2.9	4.9	4.3	2.91
Cetane No.		62	60	45	52	51	54		45–60
Flash point	°C	156.5	152	178	126	100	176	182	71.5
Cloud point	°C	10.5	16	1	3	0	5	–5	2
Oxidation[b]	h	7.5	1.6	2.1		25		15	23.7

[a]Used cooking oil; [b]Oxidation stability at 110°C

The large range in values for the CN of soybeans underpins the large range in fatty acid composition of specific oils in nature. The variation is dependent upon geographic location (soils) and climate (season). This has been well researched for the major traded commodities but is unknown for many of the exotic feedstocks. This effect is illustrated in Table 3.7 for coconut.

Further detail on the properties and growing habit of exotic feedstocks is given in the Appendix.

A Note on Algae

Algae are extensively researched to produce biofuels. The table gives typical values for the lipid composition. However, a great deal of interest is in *Botryococcus braunii*, which is the common green algae of toxic lake blooms. As well as lipids, this algae can produce large volumes of high molecular weight isoprenoid hydrocarbons, which are extraneous to the

---

[24] D.S. Kim, M. Hanifzadeh and A. Kumar, "Trend of Biodiesel Feedstock and its Impact on Biodiesel Emission Characteristics," *Environ. Prog. Sustain. Energy* **37**, 7–19 (November 2017). doi:10.1002/ep 12800.

[25] J.G. Ge, A.K. Yoon and N.J. Choi, "Using Canola Oil Biodiesel as an Alternative Fuel in Diesel Engines," *MDPI Appl. Sci.* **7**, 881 (2017).

Table 3.6. Properties of Pertinent FAME for Use as Bio-Fuel

	SN[a]	Iodine No. (estimate)	Density	Viscosity (estimated)	Distilled % at 360°C	Cetane No.	C18:0	C18:1	C18:2	C18:3	C22:1	Total
Algae	202.1	121.4	0.865	3.932	98.8	46.0	2.6	31.2	48.2	1.9	0.0	98.0
Candlenut	201.1	176.1	0.885	3.709	100.0	33.8	6.7	10.5	48.8	28.5	0.0	100.0
Canola	199.8	123.3	0.881	4.164	99.5	45.9	3.1	60.2	21.1	11.1	0.5	99.9
Coconut	270.5	9.3	0.868	2.497	100.0	64.4	2.9	6.9	1.7	0.0	0.0	20.7
Coconut	270.7	9.2	0.871	2.551	100.3	64.4	2.9	6.8	1.6	0.1	0.0	20.3
Copaiba	199.6	111.2	0.879	3.644	93.4	48.6	4.4	31.0	45.3	0.0	0.0	93.4
Corn	204.3	116.4	0.884	3.999	100.0	46.8	3.1	18.5	54.2	0.5	0.0	99.2
Corn	200.8	123.8	0.883	3.880	98.3	45.6	2.2	31.1	51.0	1.0	0.1	97.8
Cotton	204.2	112.9	0.885	3.878	99.0	47.6	2.7	18.2	52.5	0.2	0.2	97.6
Crambe	180.2	97.7	0.876	4.667	35.5	54.6	1.0	16.7	7.8	6.9	55.7	89.8
Crambe	168.9	91.6	0.875	4.250	37.5	58.0	0.7	18.9	9.0	6.9	52.5	90.0
False flax	191.6	161.6	0.885	3.368	79.0	38.4	2.5	19.5	15.5	36.0	1.0	80.0
Fish oil	177.2	125.6	0.93	1.863	64.3	48.8	6.2	17.3	4.4	0.9	0.0	51.3
Tuna	177.8	211.5	0.92	1.469	47.7	29.4	4.1	14.8	1.9	0.0	0.5	38.9
Whale	180.2	83.4	0.92	2.343	80.2	57.8	2.0	51.1	1.4	0.0	0.6	63.4
Hemp	214.5	185.0	0.886	4.263	106.5	30.1	2.5	13.0	60.0	23.5	0.0	106.5
Honge	205.1	104.9	0.877	4.259	101.0	49.3	0.0	46.0	35.0	0.0	0.0	101.0
Honge	201.8	65.9	0.889	4.353	98.0	58.5	18.0	55.0	9.0	0.0	0.0	96.0

Jatropha	179.8	104.9	0.880	3.342	89.9	53.1	6.9	43.1	34.3	1.4	0.0	89.9
Jatropha	202.6	95.0	0.887	4.227	99.6	51.9	9.7	40.8	32.1			98.2
Kapok	169.7	86.0	0.875	2.840	100.0	59.1	3.0	21.0	37.0	0.0	0.0	83.0
Linseed	201.4	201.0	0.888	3.557	100.0	28.2	2.4	19.7	18.0	54.9	0.0	100.0
Mahua	200.4	59.3	0.867	4.423	98.5	60.2	22.7	37.0	14.3	0.0	0.0	98.5
Mallee	204.8	76.8	0.872	4.327	100.0	55.7	0.0	55.0	15.0	0.0	0.0	100.0
Milk bush	189.6	71.3	0.874	3.533	90.9	59.1	7.4	47.5	14.8	0.5	0.0	90.2
Mustard	179.6	104.7	0.876	4.895	46.0	53.1	0.8	23.2	8.9	10.4	43.1	89.0
Neem	201.8	65.9	0.889	4.353	98.0	58.5	18.0	55.0	9.0	0.0	0.0	96.0
Palm kernel	260.4	18.0	0.871	2.771	100.3	63.2	2.2	15.5	2.3	0.0	0.0	28.2
Palm oil	208.5	55.1	0.876	4.354	100.0	60.1	4.4	39.6	10.3	0.3	0.0	98.7
Palm stearin	203.5	38.8	0.875	4.036	96.5	64.4	5.0	29.2	6.7	0.1	0.0	95.1
Palm olein	200.9		0.876	4.432	100.0	73.5	3.6	75.3	9.5	0.6	0.0	100.0
Palm olein	207.6	59.6	0.877	4.360	100.0	59.2	4.2	43.3	11.0	0.2	0.0	98.8
Peanut	201.0	103.6	0.877	4.274	99.8	50.1	8.9	47.1	32.9	0.5	0.2	100.0
Peanut	204.0	99.8	0.879	3.883	94.8	50.6	3.2	51.8	28.5	0.1	0.2	94.8
Poppy	201.7	154.2	0.884	3.804	100.0	38.7	2.0	11.0	72.0	5.0	0.0	100.0
Prickly acacia	195.2	101.8	0.877	3.698	95.0	51.4	7.8	28.1	38.2	2.6	0.0	94.8
Radish	181.6	107.8	0.877	4.593	49.0	52.1	2.0	18.0	11.0	13.0	37.0	86.0
Ramtil	215.3		0.900	4.467	105.1	71.7	7.0	23.2	62.1	2.0	0.0	102.9

(*Continued*)

Table 3.6. (*Continued*)

	SN[a]	Iodine No. (estimate)	Density	Viscosity (estimated)	Distilled % at 360°C	Cetane No.	C18:0	C18:1	C18:2	C18:3	C22:1	Total
Ramtil	198.6	124.1	0.880	3.916	98.4	45.9	7.9	27.0	55.0	0.0	0.0	98.4
Ramtil	200.1	146.5	0.883	3.729	98.4	40.6	6.5	6.5	76.6	0.6	0.0	98.4
Rapeseed	182.8	105.3	0.876	5.383	49.1	52.5	2.8	21.9	13.1	8.6	50.9	100.0
Rice bran	202.5	109.2	0.881	4.087	99.4	48.7	1.8	42.3	37.1	1.3	0.0	98.9
Rubber	200.4	138.7	0.880	3.910	99.4	42.3	8.7	24.6	39.6	16.3	0.0	99.4
Safflower	201.0	150.2	0.885	3.859	100.0	39.7	3.3	14.4	75.5	0.1	0.0	99.9
Safflower	196.8	95.0	0.879	4.114	97.5	52.7	2.3	73.6	15.8	0.0	0.0	97.4
Safflower	199.5	149.1	0.885	3.734	98.7	40.1	2.9	13.8	75.3	0.0	0.0	98.5
Soybean	201.6	141.2	0.882	3.924	100.0	41.6	4.7	22.5	54.1	8.3	0.0	99.9
Soybean	198.9	137.6	0.882	3.735	97.6	42.8	3.9	21.9	53.5	7.5	0.2	97.6
Sunflower	200.8	145.6	0.884	3.909	100.0	40.7	5.9	16.0	71.4	0.6	0.0	99.9
Sunflower	202.5	135.8	0.882	3.935	99.4	42.7	4.6	26.7	61.2	0.1	0.1	99.3
Tallow beef	205.2	50.3	0.898	4.600	100.0	61.6	19.2	48.9	2.7	0.5	0.0	96.5
Tallow lamb	205.3	40.2	0.905	4.714	100.0	63.8	30.5	36.0	4.3	0.0	0.0	95.4
Tallow pork	204.5	61.6	0.882	4.517	100.0	59.1	15.8	47.1	8.9	1.1	0.0	98.4
Tung	201.9	225.7	0.890	3.434	100.3	22.5	2.4	11.2	14.6	69.0	0.0	100.3

[a]Saponification number

Table 3.7. Fatty Acid Composition Variation Due Geographic Location for Coconut

	C6:0	C8:0	C10:0	C12:0	C14:0	C16:0	C18:0	C18:1	C18:2	C20:0	C20:1
Philippines	0.5	7.7	6.5	47.45	18.6	8.9	2.75	6.35	1.7	0.15	0.05
PNG	0.5	7.2	6.6	48.7	17.95	8.8	2.65	6.05	1.5	0.1	
Vanuatu	0.55	8.35	7.2	47.75	17.35	8.4	2.6	5.95	1.7	0.1	0.1
North Sulawesi	0.5	7.3	6.3	45.9	18.1	9.7	2.7	7.4	1.9	0.1	
Sri Lanka	0.45	7.35	6.05	50.95	19.1	7.45	2.9	4.9	0.75	0.1	

Table 3.8. Impact of the Use of Different Alcohols in Biodiesel Properties

Ester	Cetane	HHV MJ/kg	Viscosity cSt	Cloud point °C	Pour point °C
Methyl	46.2	39.8	4.08	2	−1
Ethyl	48.2	40.0	4.41	1	−4
Butyl	51.7	40.7	5.24	−3	−7

cells of the algae. This is harvested, essentially by solvent extraction processes, and then hydrocracked to form gasoline, jet fuel and diesel.[26]

Impact of Different Alcohols on Biodiesel Properties

Methanol is not generally considered a renewable alcohol, unlike ethanol and butanol, which can be produced by fermentation routes. Consequently, there has been some interest in using these alcohols to transesterify vegetable oils and fats instead of the commonly used methanol. A comparison is given in Table 3.8.

The data in the table show that the heavier esters (ethyl and butyl) increase cetane, higher heating value (HHV), viscosity and lower the cloud and pour point of the biodiesel.

Use of FAME as a Diesel Fuel

The conventional nomenclature for describing biodiesel/diesel blends is BXX, where the XX is the volume percent of biodiesel in the blend, for example, B20 is a blend of 20% biodiesel and 80% conventional diesel.

---

[26] Wikipedia, Wikimedia Foundation, downloaded August 15, 2024.

There are differences between biodiesel and conventional fuel. Except in specific circumstances, these differences do not manifest themselves in biodiesel/diesel blends with a biodiesel content of 5% or less. Thus, OEMs, expressed in the Worldwide Fuel Charter,[27] generally accept biodiesel/diesel blends to a level of 5% in low sulphur (<30 ppm sulphur), Category 4 diesel fuel. FAME should be absent in Category 5 fuels. Despite this, there have been moves in several jurisdictions to increase the permitted amount of biodiesel to 7.5% (B7.5) or higher for general use. Some OEMs produce vehicles that can operate with 20% FAME (B20), and a few can operate on B100.

Differences between Diesel and Biodiesel

Fundamental differences between biodiesel and petroleum diesel influence the properties of blends higher in biodiesel concentration than B5. The key differences that adversely influence the properties of the biodiesel blend are

- The production of biodiesel does not involve a distillation process, whereas (except for additives) diesel is produced from distilled hydrocarbon streams. The consequence is that the biodiesel components of the blend will contain heavy materials subject to thermal cracking in the hotter parts of the delivery system and the engine, hence potentially increasing engine fouling.
- Biodiesel contains a high level of olefins relative to diesel, which because of hydrotreatment to reduce sulphur is generally low in olefins. Olefins lead to poor thermal stability, leading to coking and poor oxidative stability, leading to inferior storage life.
- Biodiesel contains oxygen, which as well as lowering the energy content of the blend, fundamentally influences the injection and burning properties of the blend, increasing the emissions of nitrogen oxides.
- The oxygen of biodiesel results in higher solvating and detergent power than hydrocarbons. Biodiesel will absorb more water and suspend solids better than petroleum diesel.
- The generally larger molecules of biodiesel result in higher cloud points and cold filter plugging points than conventional diesel. These

---

[27] Worldwide Fuel Charter, published by ACEA, Alliance, EMA or JAMA, October 2019.

parameters set the lower operability temperature of the fuel; hence, a high cloud point fuel cannot be used at low ambient temperatures.

However, the presence of oxygen in the blend improves some properties of the blend relative to the conventional diesel used in the blend. Biodiesel has a higher viscosity and, in particular, lubricity than diesel, especially low-sulphur diesel. This improves the engine wear properties in the blend. Oxygen greatly reduces the amount of particulate matter and hydrocarbon emissions. In addition, biodiesel does not contain aromatics. As a consequence, the emitted particulate matter is considered to be more benign than the particulate matter emitted by petroleum diesel.

One way to improve the properties of biodiesel would be to hydrogenate the olefins and esters and reduce the molecules present to hydrocarbons and distil the required lighter fractions away from the longer-chain heavier fractions. However, these processes would be unduly onerous and costly for the production of biodiesel.

Use of FAME in SAF

The properties of FAME do not facilitate its use as a jet fuel or as a blendstock. At present, the limit to the use of FAME in jet fuel is 5 ppm. Nevertheless, trails have been conducted using higher levels of FAME.

Green Diesel and SAF

The issues outlined above with FAME limit the uptake of biodiesel into the diesel fuel stream. Without specially adapted engines, the limit appears to be about 10% (B10), although some jurisdictions are pushing for higher levels. The key problems stem from the fundamental properties of FAME, which arise primarily from the presence of a carboxylic ester group. A way to ameliorate the problem would be to remove this group by hydrogenation and produce a pure hydrocarbon, which would be more compatible with mineral diesel. This idea was taken up by

- conventional refinery operations that possess the hydrogenation facilities required, and

- so-called "Biorefineries", which have bespoke hydrogenation equipment to accomplish the task.

Either way, since the produced product emanated from vegetable oils and fats, the term "green diesel" is used to describe this product.

Refinery Operations to Green Diesel

The workhorse for the conversion of vegetable oils and fats (including tallow) to green diesel is the gas oil hydrotreater. In the normal operation of a refinery, this has the duty to treat gas oil to hydrogenate olefins and remove sulphur, nitrogen and oxygenates to produce a specification diesel blendstock. The outline operation is illustrated in Figure 3.8.

Heavy oils containing large amounts of unsaturated material (di-olefins etc.), such as distillate produced from coker operations, are hydrotreated in a unit prior to blending with the main gas oil feed. Vegetable oils and tallow can be blended in this feed stream. The feed is heated, mixed with hydrogen and passed to the hydrogenation reactor

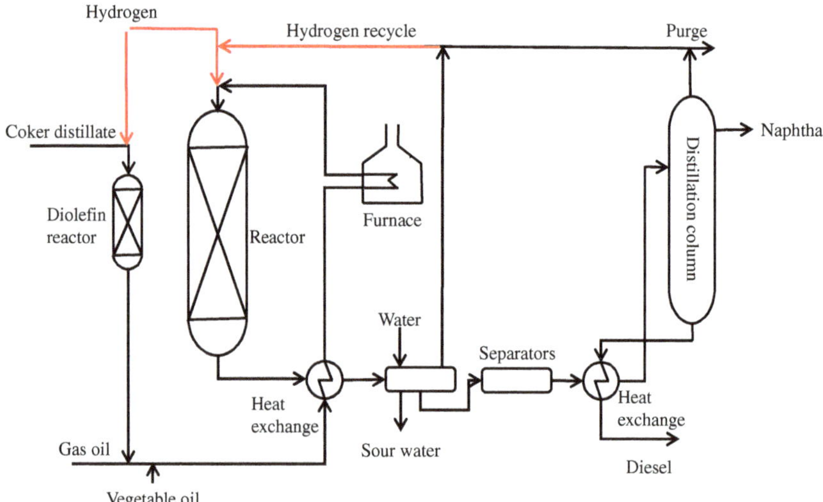

Figure 3.8.   Gas Oil Hydrotreater Operation Incorporating Vegetable Oil

operating at typically 320°C to 400°C at 6 to 10MPa pressure, charged with a catalyst that is typically cobalt/molybdenum oxides supported on alumina (Co-Mo/Al$_2$O$_3$).

For gas oil, the process

- hydrogenates the olefins present to paraffins,
- removes sulphur in the hydrocarbons, producing hydrogen sulphide,
- removes nitrogen compounds producing ammonia,
- removes oxygen compounds producing water.

For gas oil, the principal duty is to remove sulphur to less than 10 ppm in the final product.

For vegetable oils and tallow, the duty of the process is to remove the carboxylate group to produce a hydrocarbon product, carbon dioxide, water and propane (from the glycerine group). In accomplishing this, any unsaturated material in the fatty acid chains is reduced to paraffins.

There are two reaction mechanisms that run in parallel in about equal measures. The first mechanism is that of hydrogenation, which removes the oxygen in the fatty acid chains as water.

$$R-CO_2-R' + 4H_2 = R-CH_3 + H-R' + 2H_2O$$

where R is a fatty acid chain and R' is the CH_2 moiety of the glyceride group (this is reduced to propane when all three fatty acid groups are hydrogenated). Note four hydrogen molecules are required for this reduction, and an additional hydrogen molecule is required to reduce each of any olefins present in the fatty acid chain.

The second mechanism is one of decarboxylation, which eliminates the carboxylate group as carbon dioxide. This mechanism produces a carbon chain that is one carbon atom less in length than the original.

$$R-CO_2-R' + H_2 = R-H + H-R' + CO_2$$

Fatty acid chains are dominated by chains with even carbon numbers (C_{14}, C_{16}, C_{18} etc.); the decarboxylation mechanism produces corresponding chains with one less carbon – odd carbon numbers. Since both mechanisms occur in approximately equal proportions, the resulting product has a more homogeneous distribution of carbon numbers in the resulting product, and these better match that of petroleum diesel.

After the reaction, the product is cooled, and the water, the hydrogen sulphide and the excess hydrogen are separated prior to passing to a distillation column, which separates the diesel from the lighter naphtha fractions that are produced from spurious cracking reactions.

Thus, a refiner adding vegetable oils or fats to the gas oil hydrotreater will produce a green diesel/mineral diesel mixture roughly proportional to the quantum of vegetable matter added. Thus, if 5% of the carbon boiling in the diesel range is contributed by the vegetable matter, the diesel product could be regarded as equivalent to a B5 diesel.

The use of vegetable matter in a conventional gas oil hydrotreater is limited by the capacity of the equipment.[28] The processing of vegetable matter requires more hydrogen than required for conventional gas oil hydrotreating. Furthermore, the reduction reactions are quite exothermic, which compromises the capacity of the heat exchangers in the system.

Biorefineries Producing Green Diesel

In refinery operations using gasoil hydrotreaters, the volume of gas oil exceeds the volume of vegetable oil and fats being processed. The resulting product is dominated by gas oil-derived material. This reduces the impact of some of the properties of fatty acid chains, which lead to poor-quality diesel. The fatty acid chains are linear, which although delivering good cetane, are poor from the standpoint of cold flow properties.

Thus, refining of pure vegetable oils and fats results in products with unacceptable cloud point, pour point and cold temperature filtration. In a biorefinery processing entirely vegetable oils, fats and tallow, isomerisation of the hydrotreated product is required. An outline flow diagram for a biorefinery operation is illustrated in Figure 3.9.

The feed vegetable oil is pumped to a heater mixed with hydrogen and then on to the hydrotreater. The product passes to a separator, which removes the produced water and carbon dioxide. The hydrocarbon liquids

---

[28] S. Green, "Key Considerations for the Design and Operation of a Renewable Diesel Unit," *Hydrocarbon Processing*, January 2022.

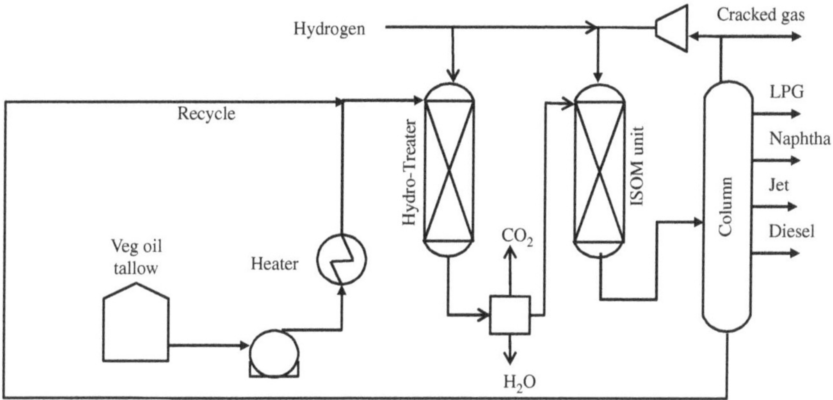

Figure 3.9.   Outline Operations in a Biorefinery

are reheated (not shown in the figure) and passed to the isomerisation unit. This has the duty to isomerise the linear hydrocarbon chains into branched chains, which although poorer in cetane have acceptable cold flow properties.

Following the isomerisation, the mixed products are distilled to produce green LPG (mainly propane produced from the glyceride group of the feedstock), green naphtha (which comprises branched paraffins lower than C_{10}), green jet fuel (SAF, branched C_{10}–C_{14} paraffins) and green diesel (branched paraffins C_{14}–C_{18}). Heavier material is recycled. For most feedstocks, green diesel is the major product.

There is a small amount of cracked gas, which is purged from the system. Hydrogen-rich gases are recycled.

Relative to FAME biodiesel, green diesel has the beneficial properties of

- commercially proven and widely practiced technology,
- green diesel is a distilled product with no insoluble materials being present,
- green diesel has a higher HHV than biodiesel; no carboxylate groups are present,
- green diesel has higher thermal stability than biodiesel since there are no olefins present.

Table 3.9. Mineral Diesel, FAME and Green Diesel Compared[29]

	Mineral diesel (US — low sulphur diesel)	FAME	Green diesel
Oxygen (wt%)	0	11	0
Density (kg/m^3)	840	880	780
Sulphur (ppm)	<10	<1	<1
HHV (MJ/kg)	43	38	44
Cloud point (°C)	−5	−5 to +15	−10 to +20
Distillation (°C)	200–350	340–355	265–320
Cetane No.	40	55–65	70–90
Stability	Good	Marginal	Good

The properties of green diesel relative to FAME and mineral diesel are illustrated in Table 3.9.

The main point to note is the low density of green diesel. This compromises the specification for fuel density in many jurisdictions, which requires a minimum density of 820 kg/m^3.

Sustainable Aviation Fuel

Because it has the potential to introduce insoluble matter, FAME is regarded as a contaminant in jet fuel and must be less than 50 ppm.[30]

The jet fuel cut represents SAF. For most feedstocks, the volume of the SAF cut is small. It can be increased by selecting feedstock high in fatty acids of the appropriate chain length. This is mainly coconut oil, which has a high level of C_{12} and C_{14} fatty acids (see Table 3.6).

Because of the limited availability of coconut oil, there has been a growing interest in rape (not Canola), which has a high concentration of erucic acid (C_{22}, C22:1). This can be selectively hydrocracked, producing a product with a carbon number around C_{11}, that is, in the SAF range.

---

[29] C. Gosling, J. Holmgren and T. Marker, "New Developments in Renewable Fuels Off More Choices," *Hydrocarbon Processing*, September 2007.

[30] Bureau Veritas: Fame in Jet Fuel, https://commodities.bureauveritas.com/oil-gas/fuel-quality-testing-monitoring/fame-jet-fuel.

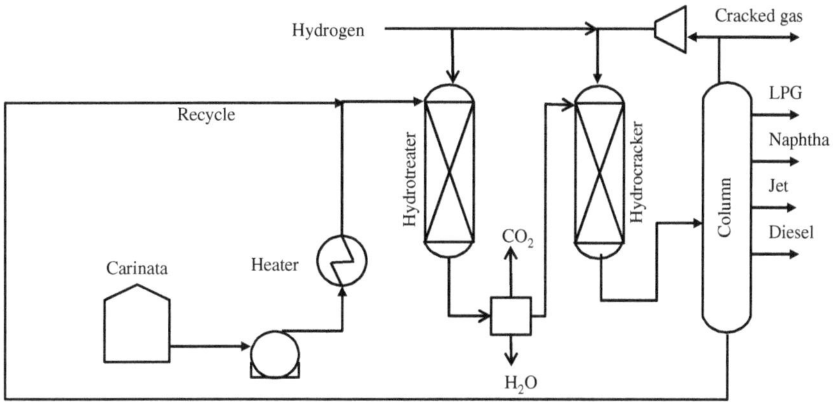

Figure 3.10. Possible Processing of Carinata to SAF

There has been extensive work on improving the yield of erucic acid in rapeseed varieties,[31] and several have come to prominence as potential SAF fuels, particularly the variety known as carinata.

In processing carinata (or similar feedstock), the isomerisation unit of the biorefinery is replaced by a hydrocracking unit. This has the duty to both selectively hydrocrack the long C_{22} chains and to isomerise the product into the branched isomers required to achieve the low cold flow properties of jet fuel (Figure 3.10).

The introduction of hydrocracking further increases the hydrogen demand of the system. It also increases the production of cracked gases exiting the unit.

It is envisaged that carinata would largely replace Canola as a bio-crop. For fuel production, this has the advantage of being an intermediary crop sown between summer and winter food crops such as wheat and thus does not directly impinge on agricultural food crop yields. This is an important consideration in some jurisdictions, such as the EU.

Algae oils could also be processed in a similar way. *B. braunii* produces large volume of isoprenoid hydrocarbons, which can be hydrocracked.[32]

---

[31] A. Jadav et al., "Production of 22:2 and 20:1 in *Brassica Carinata* and Soybean Breading Lines via Introduction of *Limnanthes Genes,*" *Mol. Breed.* **15**, 157–167 (2005).
[32] L.W. Hillen, G. Pollard and L.V. Wake, "Hydrocracking of the Oils of *Botryococcus Braunii* to Transport Fuels," *Biotechnol. Bioeng.* **24**, 193–205 (1982).

They have a typical formula of $C_{36}H_{62}$. Although not containing oxygenates, hydrocracking to produce paraffinic compounds will require large volumes of hydrogen, as

$$C_{36}H_{62} + 7H_2 = 2C_{18}H_{38}$$

So, processing issues and costs would be expected to be similar to the processing of vegetable oils.

Production Costs

Production costs are estimated by the method detailed in the Appendix.

Production Cost of Biodiesel (FAME)

The production cost of biodiesel (FAME) for several feedstocks is estimated in Table 3.10 for average 2022 prices.[33] The feedstocks are Canola, soybean, crude palm oil (CPO), tallow and used cooking oil (UCO). The estimate of the cost is based on a 100 million gallon/year production with the feedstock, methanol and crude glycerol prices taken as 2022 averages for a US operation.

The table illustrates the variation depending on the price of the various feedstocks. For the vegetable oils in this snapshot, CPO delivered the lowest price and soybean the highest production cost estimate. For 2022, the gas oil average was \$134.8/bbl. Without government subsidies (tax breaks, grants and biofuel mandates), the vegetable oils would be uncompetitive against mineral diesel fuel. Tallow-produced biodiesel is more competitive, but this product suffers from poor cold flow properties and would require further treatment.

UCO delivered biofuel well below the price for virgin feedstocks. The price for this snapshot is taken at 50% of that for Canola. This is somewhat arbitrary since the UCO price is a function of the collector's cost to obtain it; the cost could be lower than indicated for some operations. Since the availability of UCO is limited, an estimate for a smaller facility was evaluated and, *prima facie*, shows potentially a good margin for a smaller

---

[33] Data based on Lurgi AG process in *Hydrocarbon Processing,* Refining Processes 2006.

Table 3.10.　Estimates for the Cost of Producing Biodiesel (FAME) from Various Sources

Feedstock		Canola	Soybean	CPO	Tallow	UCO	UCO
Feedstock density		0.9155	0.9170	0.9040	0.9040	0.9155	0.9155
Feedstock molecular weight		876.6	930.0	854.4	872.3	876.0	876.0
Capacity (product)	MMgal/y[b]	100.00	100.00	100.00	100.00	100.00	50.00
	ML/y	378.60	378.60	378.60	378.60	378.60	189.30
FAME density	kg/L	0.860	0.870	0.870	0.870	0.860	0.860
FAME	t/y	325,596	329,382	329,382	329,382	325,596	162,798
ISBL CAPEX 2018	M$	44.21	44.21	44.21	44.21	44.21	27.21
CEPI 2022		1.361	1.361	1.361	1.361	1.361	1.361
ISBL % PLANT		0.65	0.65	0.65	0.65	0.65	0.65
Total Capex 2022		92.57	92.57	92.57	92.57	92.57	56.98
Return on capital[a]		13.84%	13.84%	13.84%	13.84%	13.84%	13.84%
Return on capital	$M/y	12.81	12.81	12.81	12.81	12.81	7.89
Capital cost	c/L	3.38	3.38	3.38	3.38	3.38	4.17
	$/bbl	5.38	5.38	5.38	5.38	5.38	6.62
Feedstock	t/y	325,596	329,382	329,382	329,382	325,596	162,798
Caustic soda	t/y	1302.4	1317.5	1317.5	1317.5	1302.4	651.2
Methanol	t/y	32,559.60	32,938.20	32,938.20	32,938.20	32,559.60	16,279.80
Methanol price	$/t	400.00	400.00	400.00	400.00	400.00	400.00
Operating cost							
Caustic (@$740/t)	$M/y	0.39	0.40	0.40	0.40	0.39	0.20
Methanol	$M/y	13.02	13.18	13.18	13.18	13.02	6.51

(*Continued*)

Table 3.10. *(Continued)*

Feedstock		Canola	Soybean	CPO	Tallow	UCO	UCO
Labour (3% Capex/y)	$M/y	2.78	2.78	2.78	2.78	2.78	1.71
Maintenance (3.5% Capex/y)	$M/y	3.24	3.24	3.24	3.24	3.24	1.99
Other (1.5% Capex)	$M/y	1.39	1.39	1.39	1.39	1.39	0.85
Operating cost	$M/y	20.82	20.98	20.98	20.98	20.82	11.27
	c/L	5.50	5.54	5.54	5.54	5.50	5.95
	$/bbl	8.74	8.80	8.80	8.80	8.74	9.46
Feedstock	$/t	1,604.00	1,667.00	1,276.00	1,073.51	328.28	802.00
Feedstock costs	$M/y	522.26	549.08	420.29	353.59	261.13	130.56
Feedstock costs	c/L	137.94	145.03	111.01	93.40	68.97	68.97
	$/bbl	219.19	230.45	176.40	148.41	109.60	109.60
Subtotal costs	$M/y	555.89	582.87	454.08	387.38	294.76	149.72
	c/L	146.83	153.95	119.94	102.32	77.86	79.09
	$/bbl	233.31	244.63	190.58	162.59	123.71	125.67
Glycerin	t/y	41,676.29	42,160.90	42,160.90	42,160.90	41,676.29	20,838.14
Price	$/t	300.00	300.00	300.00	300.00	300.00	300.00
Credits	$M/y	12.50	12.65	12.65	12.65	12.50	6.25
	c/L	3.30	3.34	3.34	3.34	3.30	3.30
Net production cost	$M/y	543.38	570.22	441.43	374.73	282.26	143.46
	c/L	143.52	150.61	116.60	98.98	74.55	75.79
	$/bbl	228.06	239.32	185.27	157.28	118.46	120.43

[a](2-Year construction; 20-year life, 10% DCF, 1% royalty); [b]Million gallons per year.

operation. In this snapshot, UCO biodiesel was below the cost of mineral diesel (as gas oil).

In this analysis, the feedstock cost typically represents 95% of the production costs, which emphasises the importance of facilities being feedstock flexible to take advantage of temporary lows in specific feedstocks.

Production Cost of Green Diesel — Case Studies

Three cases are evaluated for feedstocks: tallow, Canola and CPO. Unlike FAME, the process requires hydrogenation and decarboxylation of the fatty acid groups of the vegetable oil. Methanol and caustic are not required, rather, a large amount of hydrogen is required to

- hydrogenate the olefinic bonds; hydrogen requirement will depend on the degree of unsaturation in the fatty acid chains of the vegetable oil,
- hydrogenate the glycerine ester group to form a hydrocarbon chain and propane,
- hydrogenate approximately half of the oxygen in carboxylate groups of fatty acid to water,
- provide hydrogen for the decarboxylation mechanism to produce the shorter hydrocarbon chain and propane,
- provide hydrogen to minimise catalyst fouling,
- provide hydrogen for isomerisation of the product hydrocarbon chains,
- provide hydrogen for spurious (unwanted) cracking reactions.

Most of the hydrogen is required for the first four reactions. In theory, hydrogen requirements for catalyst fouling and cracking should be zero; in practice, there are some losses. Optimum performance is obtained by minimising cracking reactions, but this is difficult to achieve in practice. Cracking will produce some light gas products (mainly methane).

The econometric analysis is informed by the data available for the Neste Biorefinery in Singapore and UOP Ecofining™ Process.[34]

---

[34] Honeywell, *Reimaging Refining with Ecofining™*, Honeywell International Inc., 2025, https://uop.honeywell.com/en/industry-solutions/renewable-fuels/ecofining.

Table 3.11. Statistics for Hydrogenation of Tallow to Green Diesel

Feedstock	kt/y	1,000.0
Products		
LPG	kt/y	54.13
Hydrocarbons	kt/y	831.28
Water	kt/y	62.50
Carbon dioxide	kt/y	76.39
Cracking of products	mass %	5%
Gasoline	ML/y	63
Gasoline	bbl/d	1,166
Diesel	ML/y	1,018
Diesel	bbl/d	18,840

The simulation uses tallow as the feedstock. Salient details are given in Table 3.11. It is assumed that the ratio between hydrogenation and decarboxylation is 1:1.

The estimated production cost for the hydrocarbon products, mainly green diesel (94% of the liquid products), is given in Table 3.12.

Again, the estimated costs of production are considerably higher than for mineral diesel ($135/bbl, 2022 average), and again the results show a marked variation in the cost of production with feedstock. This process fully hydrogenates the fatty acid chains and concomitantly isomerises them to branched isomers, thus counteracting the negatives of poor cold flow properties for feedstocks such as tallow. For this snapshot, the lowest production cost was for tallow, with Canola the highest, reflecting the costs of the feedstock oil or fat.

Compared to FAME, the production costs are higher, but the green diesel product has the advantage of a "drop-in" fuel and can be used in any combination with mineral diesel. Its only drawback is its low density, but blending with higher-density mineral diesel will easily overcome any issues arising from this.

Table 3.12. Estimated Production Cost of Green Diesel from Tallow, Canola and Crude Palm Oil

Feedstock		Tallow	Canola	CPO
Feedstock density		0.9040	0.9155	0.9040
Feedstock molecular weight		872.3	876.6	854.4
Capacity (product)	ML/y	1,025.64	1,025.64	1,025.64
FAME density	kg/L	0.780	0.780	0.780
FAME	t/y	800,000	800,000	800,000
ISBL CAPEX 2018	M$	770.00	770.00	770.00
CEPI 2022		1.361	1.361	1.361
ISBL % Plant		60%	60%	60%
Total Capex 2022		1,047.97	1,047.97	1,047.97
Return on capital[a]		13.84%	13.84%	13.84%
ROC	$M/y	145.04	145.04	145.04
Capital costs	c/L	14.14	14.14	14.14
	$/bbl	22.47	22.47	22.47
Feedstock	t/y	977,320	977,320	977,320
Hydrogen	t/y	36.25	36.25	36.25
Hydrogen price	$/t	3,000.00	3,000.00	3,000.00
Operating costs				
Hydrogen cost	$M/y	0.11	0.11	0.11
Labour (3% Capex)	$M/y	31.44	31.44	31.44
Maintenance (3.5% Capex)	$M/y	36.68	36.68	36.68
Other (1.5% Capex)	$M/y	15.72	15.72	15.72
Operating cost	$M/y	83.95	83.95	83.95
	c/L	8.18	8.18	8.18
	$/bbl	13.01	13.01	13.01
Feedstock	$/t	1,073.51	1,604.00	1,276.00
Feedstock costs	$M/y	1,049.16	1,567.62	1,247.06
Feedstock costs	c/L	102.29	152.84	121.59
	$/bbl	162.54	242.87	193.20

(Continued)

Table 3.12.   (*Continued*)

Feedstock		Tallow	Canola	CPO
Subtotal costs	$M/y	1,278.15	1,796.61	1,476.05
	c/L	124.62	175.17	143.91
	$/bbl	198.02	278.34	228.68
Propane	t/y	54,132	541,326	54,132
Price	$/t	400.00	400.00	400.00
Credits	$M/y	21.65	21.65	21.65
	c/L	2.11	2.11	2.11
Net production cost	$M/y	1,256.49	1,774.95	1,454.39
	c/L	122.51	173.06	141.80
	$/bbl	194.67	274.99	225.33

[a]2-Year construction; 20-year life, 10% DCF, 1% royalty

Because it is a drop-in fuel, in most jurisdictions, a producer can supply a specific end user, such as a specific transport fleet, and if the end user agrees the density standard can be waived.

Production Cost of Sustainable Aviation Fuel

The hydrogenation/isomerisation process can be used to produce SAF. The yield of the jet-fuel cut (approximately C_{11}–C_{14}) will be low using the main feedstocks to produce green diesel. However, coconut and similar oils have a high content of C_{12} fatty acid, and if this is used or incorporated into the feedstock, the yield of jet fuel (SAF) will be higher. The cost of production will be similar to those outlined above. However, the availability of coconut oil is limited.

For higher SAF yields, the process requires the addition of a cracking function in the second (isomerisation) reactor and the use of feedstock containing large amounts of C_{22} or higher fatty acid, which can be hydrogenated, isomerised and cracked into jet-fuel components. One such feedstock is carinata, which is evaluated (Table 3.13).

Table 3.13. Estimated Cost of Production of SAF from Carinata

Density	kg/L	0.9180
Molecular wt.		973.0
Feedstock	kt/y	1,000.00
Liquid product	kt/y	868.83
ISBL CAPEX 2018 (ISBL)	M$	770.00
CEPI 2022		1.36
Capital cost 2022		1047.97
ROC % (2-year construction; 20-year life, 10% DCF, 1% royalty)	% Capex	13.84%
Return on capital	$M/y	145.04
Feedstock	t/y	1,000,000
Hydrogen	t/y	57.49
Hydrogen price	$/t	3,000.00
Operating costs		
Hydrogen	$M/y	0.17
Labour (3% Capex)	$M/y	36.68
Maintenance (3.5% Capex)	$M/y	36.68
Other (1.5% Capex)	$M/y	15.72
Operating costs	$M/y	89.25
Feedstock costs	$/t	1,604.00
Feedstock costs	$M/y	1,604.00
Subtotal costs	$M/y	1,838.29
Propane	t/y	48,529.82
Price	$/t	400.00
Credits	$M/y	19.41
Net production cost	$M/y	1,818.88
Tail gas (methane)	kt/y	10,000.00
	MGJ/y	550,000
	$/GJ	5
Tail gas revenue	M$/y	2.75

(Continued)

Table 3.13.　(*Continued*)

Gasoline	ML/y	594.79
Jet fuel	ML/y	579.71
Diesel	ML/y	0.00
		1,174.51
Cost of production	$/L	1.546
	$/bbl	245.71

Again, the evaluation is modelled on a 50/50 split of hydrogenation to decarboxylation of the fatty acid chains. For this case, the consumption of hydrogen is higher due to

- higher level of olefins in the fatty acid chains of carinata
- the requirement to crack the fatty acid chains
- spurious over-cracking reactions producing methane.

The cracking process produces an excess of tail gas (methane) in the process. As this crop largely displaces Canola as an intermediate crop between winter and summer grain crops, the cost of carinata is assumed to be the same as Canola. The production cost is estimated at $245/bbl compared to the cost of mineral jet at $127/bbl (2022).

Again, the cost of production is dominated by the cost of the feedstock.

Cost of Hydrogen

These last cases producing green diesel and SAF require hydrogen. A cost of $3,000/t is placed on this cost, which broadly reflects the cost in refinery operations. The use of "green" hydrogen has been proposed for this duty. The cost here is related to the prevailing price of electricity and generally comes in considerably higher than the $3k/t used here.[35]

---

[35] See D. Seddon, *The Hydrogen Economy — Fundamentals, Technology, Economics*, World Scientific, 2022.

Use as Jet Fuel

From the discussion of green diesel above, it will be realised that the jet fuel produced by carinata and similar crops will meet all of the requirements for Jet-A except the density criterion. Like the jet cuts produced from ethylene, this product will find its role as a blendstock with higher density kerosene to produce an acceptable Jet-A. Since a low density impacts the range achieved by an aircraft rather than any other material's impact on engine performance, some end users may be willing to use fuel produced in this manner.

4

PRODUCTION OF FUELS AND SAF FROM BIOMASS VIA GASIFICATION

Biomass

The previous chapters have illustrated the high cost of producing transport fuels from vegetable sources relative to the production of transport fuels from crude oil. The fundamental issue is the cost of feedstock, which, in one way or another, always responds to the prevailing price of crude oil or its derivatives, except in very occasional and transitory circumstances (e.g. wine gluts to produce ethanol), resulting in production costs significantly higher than the corresponding product produced from mineral oil.

To overcome this quandary, routes to sustainable fuels have been sourced from feedstock, which appear to be largely independent and lower in cost (on a carbon basis) than crude oil or its derivatives. The focus here is biomass which, in many instances, is produced as an unwanted by-product of another crop. These sources are regarded as sustainable and result in zero net carbon emissions. The term "biomass" covers a wide range of potential feedstock.

- Municipal and household waste: The use of landfill to dispose urban waste is an increasing problem, and its use as a feedstock would benefit from avoiding the costs of landfill disposal, that is, this waste stream may have a negative value. The key issue is its limited supply and variability in content, which hinders process operations to utilise it. Many jurisdictions regard urban waste as a renewable fuel, allowing it to attract government grants and subsidies.

- Bagasse: This is the waste (stalks) product from the production of sugar from sugarcane. Bagasse is relatively consistent in quality, but availability is an issue, especially since most bagasse is consumed as a power source for sugar mills in the first stage of sugar refining.
- Corn stover: This is the waste (stalks) product from the production of maize. It is like bagasse in which it could be used to power fermentation plants for the production of ethanol but is generally in excess of production. It is typically ploughed back into the cropping land, where its benefits would have to be replaced when used as an energy feedstock.
- Prairie grass: Excessive natural production from grassland (prairies) could be cropped from appropriate lands. The main issue is the cost of collection and transport to a downstream facility.
- Forestry waste: This is the waste from conventional timber harvesting producing wood-based products — housing timber or woodchips for paper manufacture and the like. In clear-felling operations, the waste timber is often left to rot or burnt *in situ*. The main issue is the cost of collection and transport to a downstream facility.
- Forest plantations: This is distinguished from forestry waste as the whole plantation is harvested for the downstream operation. This is now widely established to produce woodchips or wood pellets for power generation in North America and northern Europe. The wood pellets are used as a feedstock in converted coal-fired power stations to reduce carbon emissions. The use of plantation timber as feedstock for sustainable fuel thus competes with power generation, which will impact on its price.

Biomass Logistics

The basic method for producing sustainable fuels from these diverse materials is by gasification, hydrogasification, pyrolysis or hydropyrolysis.

- Gasification (also known as partial oxidation) is the combustion of the biomass with a restricted supply of oxygen. The objective is to maximise the production of a mixture of hydrogen and carbon monoxide, known as synthesis gas (SYNGAS), as opposed to water and carbon dioxide, the products of complete combustion.

- Hydrogasification is similar to gasification with the addition of hydrogen into the gasifier. The hydrogen may be produced *in situ* or as an additional feed stream.
- Pyrolysis is the heating of biomass in the absence of oxygen to generate a gaseous product, a liquid and a char. The objective is to maximise the liquid yield for subsequent refining to fuels.
- Hydropyrolysis is the same as pyrolysis with the addition of hydrogen into the process. This tends to increase liquid yields and improve the liquid quality before upgrading.

The principal problem with biomass gasification is the nature of the feedstock. Raw (as received) biomass has a very low specific energy, typically only 5 to 6 GJ/t for municipal waste. This is a consequence of the inherently high water (moisture) content of material as received at the facility. Drying of the "as-received" biomass will lift the specific energy to about 12 GJ/t, but this is hardly sufficient to maintain the smooth operation of a gasifier, and in many operations, a supplementary fuel is added to raise the fuel's specific energy higher. Adding natural gas is an obvious choice, and this is practiced in many large facilities for the gasification or incineration of urban waste; however, this defeats the object of producing a fossil carbon emission-free product.

Furthermore, the molecules that compose biomass contain a lot of oxygen. Biomass is generally based on a poly-cellulose matter that has an oxygen content of 55%. This bound oxygen cannot be removed before gasification or pyrolysis.

One practical problem is storage. Biomass, as-received or dry, can undergo putrefaction and spontaneous combustion brought about by bacteria and the like in the biomass (compost heap combustion is familiar to most gardeners and horticulturalists); the prevention of this increases costs.

Another issue is the variability of the biomass quality (specific energy); even dedicated biomass varies according to season. This variability can influence the downstream performance of the gasifiers and pyrolysis units. The variability is not only due to water content but also due to the nature of the material making up the biomass, which varies according to season. In addition, dedicated biomass crops (prairie grass and forest waste) can be exposed to total loss through wildfires and the like.

Many studies for bio-gasification assume that the biomass is available for free. This is not so. The use of biomass will inevitably lead to loss of soil carbon and essential minerals necessary for healthy growth. If the biomass is produced as a by-product to a cash crop, then the farmer (or forester) would want some form of compensation for this.

Finally, the collection costs of transporting the biomass to a gasifier or pyrolysis operation will be significant (typically $5–$15/t). The logistics of transporting the biomass will incur some carbon emissions from fuel used in the transport logistics. The cost and transport logistics will place limits on the scale of a biomass gasifier or pyrolysis unit. As the distance from the biomass source rises, so does the cost of biomass transport. This compromise limits the scale of bio-gasifiers/pyrolysis units in comparison to the large-scale gasification operations for coal. That is, the full economy of scale is not realised for biomass operations.

Biomass Quality

Zhou et al.[1] have compiled proximate and ultimate[2] analysis of materials constituting municipal waste, some of which are presented in Table 4.1 along with the analysis for corn stover and prairie grass. These are compared to lignite and a typical black coal used for thermal power generation.

As-received municipal waste has a specific energy of typically 5 to 6 GJ/t. After drying, the remaining vegetable matter, bone and paper have specific energies of about 15 to 17 GJ/t. Dry wood has a specific energy of nearly 20 GJ/t. On a dry ash-free (DAF) basis, these materials typically have 55% carbon plus hydrogen. The exception is bone matter that has over 65% carbon plus hydrogen. The main other element is oxygen that

---

[1] H. Zhou, A. Meng, Y. Long, Q. Li and Y. Zhang, An overview of characteristics of municipal solid waste fuel in China: Physical, chemical composition and heating value, *J. Air Waste Manag. Assoc.* **6**, 597–616 (2014).
[2] Proximate analysis and ultimate analysis are standard analytical procedures for combustion fuels.

Table 4.1. Analysis for Some Biomass Materials of Interest

	Specific energy (GJ/t)	Water (%)	Ash (%)	C (%)	H (%)	O (%)	N (%)	S (%)	
Vegetable	16.8			44.9	5.5	45.4	3.6	0.6	DAF[a]
Bone	15.7			58.0	7.2	25.4	8.7	0.74	DAF
Paper	15.1			45.5	6.3	47.7	0.2	0.2	DAF
Dry wood	19.6			50.5	5.9	43.4	0.11	0.03	Pine – DAF
Wood	12.6	28.6	28.6	37.8	4.5	28.4	0.07	0.0	As received
Corn stover	12.6	20.0	6.0	35.0	4.4	34.1	0.5	0.01	As received
Prairie grass	14.6	15.0	6.2	39.5	4.8	33.8	0.7	0.08	As received
Lignite	15.6	64.7	7.4	18.7	1.3	7.7	0.14	0.07	Victoria
Black coal	25.3	8.7	13.5	60.6	3.8	11.6	1.33	0.44	Queensland

[a]Dry ash free

can be over 40% of the total. Vegetable matter and bone have relatively high levels of nitrogen and sulphur.

As-received biomass such as wood, corn stover or prairie grass has a high moisture content. This encourages bacterial attack in storage and degrades the efficiency of gasifiers and pyrolysis units. For optimum performance, this moisture is removed before storage and gasification/pyrolysis, typically by steam driers using excess steam produced in the gasification/pyrolysis process.

The moisture content of as-received biomass is generally lower than lignite (brown coal), a widely used fossil fuel for power generation and gasification. Lignite from the La Trobe Valley in Victoria, Australia, has a moisture content over 60% and is used as a power generation fuel. Lignite used to fuel the Dakota Coal Gasification Company operation has a moisture content of about 25% to 40%. However, lignite has lower fixed oxygen content than biomass, giving them a higher calorific value measured on a DAF basis. Lignite and other poor-quality coals have been widely studied for conversion into synthetic fuels by pyrolysis and hydropyrolysis, and data on these applications can be used to inform the outcome of pyrolysis and hydropyrolysis of biomass.

Compared to biomass and lignite, black coal has a much lower as-received moisture content and lower fixed oxygen, which gives it a high specific energy. The value in Table 4.1 is for a typical export-quality thermal coal from Queensland, Australia.

Gasification of Biomass

The objective of biomass gasification is to produce SYNGAS, which is a mixture of hydrogen and carbon monoxide, in any proportion. In a secondary step, a water–gas–shift (WGS) process converts the raw SYNGAS into a gas of the required stoichiometry, which is the ratio of hydrogen to carbon monoxide in the gas to produce sustainable fuels.

In the process, the solid feed is burnt in a stream of oxygen (or, in some cases, air) and steam. The process is conducted in a gasifier of which there are many types. The chemistry of combustion is complex. The main processes involved within the gasifier are as follows:

Solid Gas Reactions

Combustion: This reaction is highly exothermic and provides the heat for the process.

$$C + O_2 = CO_2 \qquad \Delta H^\circ_{298} \; -394 \text{ kJ/mol.}$$

Steam–carbon reaction: This reaction converts water (steam) into hydrogen.

$$C + H_2O = CO + H_2 \qquad \Delta H^\circ_{298} \; +131 \text{ kJ/mol.}$$

Hydrogasification: This reaction consumes hydrogen and forms methane.

$$C + 2H_2 = CH_4.$$

The Boudouard reaction: This converts carbon dioxide into carbon monoxide. This reaction is reversible and can cause the deposition of carbon in the cooler parts of the process plant remote from the gasifier.

$$C + CO_2 = 2CO \qquad \Delta H^\circ_{298} \; +172 \text{ kJ/mol.}$$

Gas Phase Reactions

The WGS reaction: A key reaction for making hydrogen from solids. It converts carbon monoxide and steam into carbon dioxide and hydrogen. This reaction is reversible.

$$CO + H_2O = CO_2 + H_2 \qquad \Delta H^\circ_{298}\ -41.1\ \text{kJ/mol.}$$

Methanation: This process consumes hydrogen to produce methane. This reaction is reversible, with the reverse process being steam methane reforming (SMR):

$$CO + 3H_2 = CH_4 + H_2O.$$

Process Optimisation

The extent of all these processes is determined by the thermodynamics pertaining to the operational temperature and pressure of the gasifier for any given feedstock, oxidant (oxygen or air) and steam injection. The composition of the resulting SYNGAS can be estimated, that is, the relative amounts of hydrogen, carbon monoxide, carbon dioxide, water and methane determined by solving the pertinent thermodynamic equations.[3] For maximum process efficiency, the quantum of carbon monoxide plus hydrogen in the SYNGAS should be maximised. The process is optimised by changing the composition of the input stream and the operating temperature and pressure of the gasifier. For example, increasing the oxygen input will increase the gasifier operating temperature, and this will change the relative amounts of the products in the SYNGAS. Different types of gasifiers have differing optimum operating conditions.

It should be noted (Table 4.1) that biomass contains high quantities of oxygen and water. This lowers the efficiency of the gasifier by lowering the operating temperature. Adding more external oxygen to raise the reaction heat produced converts more of the carbon present in the biomass to

---

[3] C.W. Montgomery, E.B. Weinberger and D.S. Hoffman, Thermodynamics and Stoichiometry of Synthesis Gas Production, *Ind. Eng. Chem.* **40**(4), 601–607 (1948).

carbon dioxide, and excess oxygen results in the preferential burning of hydrogen:

$$2H_2 + O_2 = 2H_2O.$$

Pyrolysis Within the Gasifier

In pyrolysis processes, large molecules are cracked (pyrolysed) or hydro-pyrolysed into smaller molecules. The resultant products can undergo gasification to the desired products, but if this does not occur, because, for instance, the temperature is too low, pyrolysis tars and liquids are produced. This fouls downstream plants if not removed from the product SYNGAS stream.

The production of tars and liquids should be avoided if possible. The process is associated with gasification at relatively low temperatures of 600°C or so as opposed to temperatures of more than 900°C or higher for optimum efficiency. The problem is a particular issue with feed containing high levels of water and/or oxygen, such as lignite, biomass and municipal waste streams (MWSs).

Pure pyrolysis (i.e. no added oxygen or air) is discussed later.

Process Flow Sheet for Biomass Gasification

The process flow sheet to produce SYNGAS from biomass is similar to that for coal gasification as practiced in South Africa.[4] The approach to sustainable aviation fuel (SAF) involves two steps. The first step is the gasification of the biomass to produce a SYNGAS comprising hydrogen and carbon monoxide; the raw gas is subject to the WGS process and acid gas removal to adjust the stoichiometric ratio (H_2/CO) to an optimum value. The second step is either the conversion of the SYNGAS to methanol and then olefins or via the Fischer–Tropsch process (FTP) to a synthetic crude oil. These intermediates are then converted to SAF (Figure 4.1).

---

[4] M.E. Dry, "The Fischer-Tropsch Synthesis," in *Catalysis Science and Technology* (J. E. Anderson and M. Boudart eds.) Vol 1, Springer-Verlag, 1981.

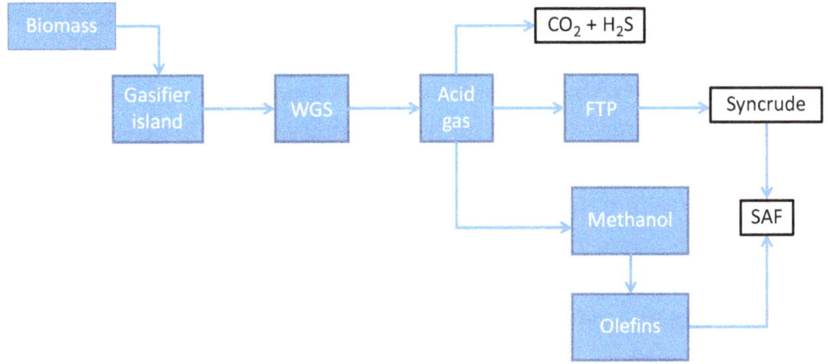

Figure 4.1. Process Flow for Biomass Gasification to Fuels Including SAF

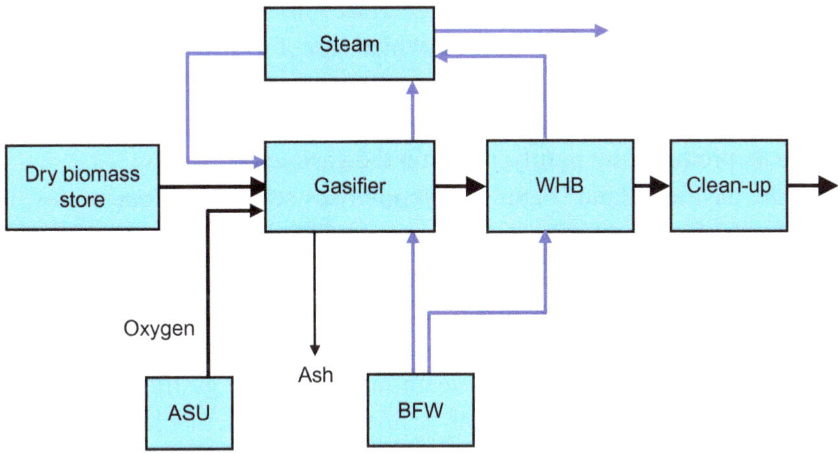

Figure 4.2. Gasifier Island for Biomass Gasification

The gasification involves several unit operations that are grouped together and commonly referred to as the gasifier island. The gasifier island shown in more detail in Figure 4.2 converts the biomass into SYNGAS (hydrogen and carbon monoxide plus steam and carbon dioxide). The downstream high-temperature WGS process converts carbon monoxide into more hydrogen and delivers the SYNGAS of the required stoichiometric ratio. For methanol synthesis, some of the carbon oxides can be as carbon dioxide; this improves the rate of reaction over the standard copper/zinc/alumina catalyst.

For FTP, the stoichiometric ratio should be about 2/1, for hydrogen/carbon monoxide with minimal carbon dioxide.

The raw SYNGAS is passed to the gas cleaning section of the plant with the duty to remove excess carbon dioxide and any hydrogen sulphide present. Any sulphur in the biomass is converted in the gasifier into hydrogen sulphide (with a small amount of carbonyl sulphide). Sulphur compounds act as poisons to methanol synthesis catalysts and Fischer–Tropsch catalysts. Any hydrogen sulphide and carbon dioxide extracted in the acid gas plant could be available for geo-sequestration or sulphur recovery.

Because biomass is considered a renewable fuel, carbon dioxide would not necessarily require disposal by geo-sequestration or carbon credit purchase but would potentially generate carbon credits if sequestered. However, many approaches embrace mixing the biomass with a fossil fuel in a co-firing operation in which case, for zero carbon emission, a disposal mechanism for some or all the carbon dioxide would be required. The clean-up section also has the duty to remove liquid and tar products produced by pyrolysis within the gasifier.

The gasifier island (Figure 4.2) comprises several unit operations. It may be the best practice that the biomass is pre-dried and stored with an inert gas environment. This may be avoided if the storage time is minimal and the gasifier is of a type that it can incorporate the drying duty.

Overall, gasifiers are net energy producers. Most designs incorporate a hollow shell for the body for steam raising, or the gasifier has steam-generating cooling coils within the gasifier. These generate high-pressure, high-temperature process steam, which is well in excess of the steam demand for other operations in the gasifier island.

There have been many approaches to the design of gasifiers for biomass.[5] The properties of biomass (high water content, high fixed oxygen content and suitable scale of operation) reduce the selection of the types available. The gasifier designs based on the moving bed gasifier[6] and the

---

[5] K.R.Craig and M.K. Mann, NREL, "Cost and Performance Analysis of Three Integrated Biomass Gasification Combined Cycle Power Systems", US DOE DE-AL36-83CH10093, October, 1996.

[6] A.A.C.M. Beenackers, "Biomass Gasification in Moving beds, a Review of European Technologies," *Renew. Energy* **16**, 1180 (January–April, 1999).

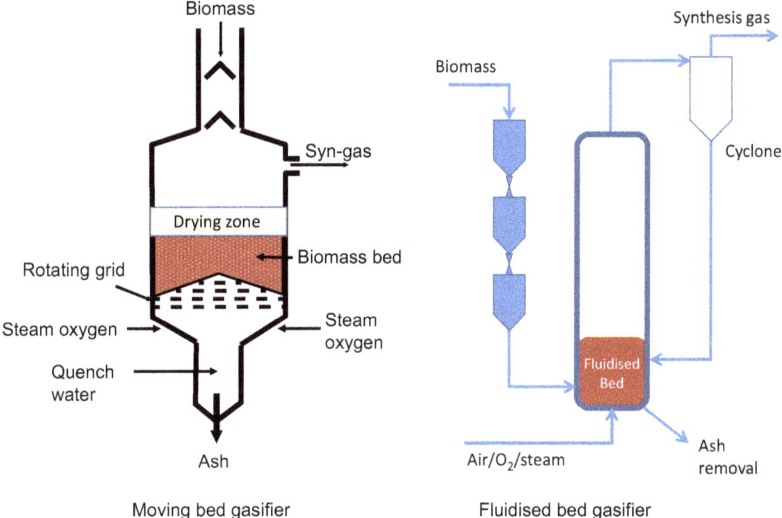

Figure 4.3. Moving Bed and Fluidised Bed Gasifiers for Biomass Feedstock

fluidised bed gasifier have proven popular for biomass[7] — they are relatively small and well proven for this type of feedstock (Figure 4.3).

In the moving bed gasifier, the design can accommodate a drying section above the gasifier bed. This performs the duty of removing moisture not removed by biomass pre-treatment before gasification. After the drying zone, the biomass moves downwards (hence the moving bed gasifier) into the combustion zone where the gasification takes place. A rotating grid maintains the position of the burning bed and allows ash or slag to fall into a collection hopper. The gasification is brought about by oxygen and steam fed to the gasifier below the rotating grid. A downside of the moving bed gasifier is the zone between drying and combustion promotes pyrolysis reactions, which contaminate the synthesis gas with liquid and tar products.

The fluidised-bed type gasifier has been proposed for biomass. This requires milling and sizing of the biomass. In this case, to avoid excessive

---

[7]Note that entrained-bed gasifiers will deliver biomass gasification. However, these gasifiers, although very efficient, are large units and would require large quantities of biomass to be economically viable.

burning of feedstock in the gasifier (to generate the heat required to remove the moisture), this type of gasifier benefits from pre-drying the raw biomass. This can be done by using steam generated by the gasifier. Above the fluidised bed is a large disengagement zone to allow unburnt material to fall back into the bed. This zone facilitates the inclusion of steam-raising tubes within the gasifier. A drawback of the fluidised bed gasifier is that there is a finite chance of biomass entering the fluidised bed and exiting the gasifier with the ash removal system. This limits the extent of combustion and contaminates the ash with non-combusted biomass. This degrades the ash for certain uses.

One of the major problems with biomass gasification is that the fixed oxygen content and consequential low specific energy result in a relatively low temperature operation, even with pure oxygen rather than air. The oxygen is supplied by an air separation unit (ASU). The low temperature operation produces excessive tars and pyrolysis liquids in the SYNGAS so that clean-up cannot be simply achieved with candle filter systems. Washing the raw gas with a solvent (such as methanol) is widely recommended. One of these processes (Rectisol™) can handle the full range of impurities found in the raw SYNGAS and can be used to extract carbon dioxide and hydrogen sulphide, that is, act as the acid–gas plant. Unfortunately, it is capital intensive and operationally complex.

As noted earlier, biomass fuel is not the best fuel for a gasifier, and its high water content results in lower-temperature operation as well as the production of liquids and tar by-products. The data given by Kreutz *et al.*[8] has been used to estimate the efficiency of a gasifier processing 1 million tonnes of corn stover. The data indicates that the production of a raw SYNGAS exiting the gasifier proceeds at a thermal efficiency of about 76%. This is considerably lower than the efficiencies of gasifiers operating on high-quality coal, which are in the region of 90%.

Once the gasifier has produced a SYNGAS of the required stoichiometric ratio and quality (impurities removed), it is passed to downstream conversion processes, which can lead to the production of green fuels and SAF. The two dominant processes of interest are the conversion of SYNGAS into SAF via methanol and the Fischer-Tropsch Process.

---

[8]T.G. Kreutz, E.D. Larson, G. Lui and R.H. Williams, *"Fischer-Tropsch from Coal and Biomass"*, 25th Annual Pittsburgh Coal Conference, 2008.

Methanol

The synthesis of methanol from SYNGAS is commercially practiced around the world. Methanol itself is a widely used intermediate for the production of formaldehyde (for formaldehyde urea resins), acetic acid and the gasoline additive methyl-*tertiary*-butyl ether (MTBE). More recently, it has been used to produce olefins.

The use of biomass via gasification[9] into SYNGAS to produce methanol will produce "green methanol".

For methanol synthesis, a SYNGAS containing some carbon dioxide is acceptable so that a certain quantum can be left in the gas. Figure 4.4 illustrates the route from SYNGAS, which is typically tailored to a stoichiometric ratio of 2/1 (H_2/CO) with about 3% to 4% carbon dioxide left in the feed gas.

Sulphur is detrimental to the methanol synthesis, and trace amounts of sulphur are removed using zinc oxide before synthesis. After the production of SYNGAS, the methanol synthesis requires compression to about 100 bar. The methanol synthesis loop comprises a reactor, a separator and recompression of the recycle gas. A purge gas can be used to produce power supplemented by steam raised in the methanol reactor and the gasifier. The crude methanol produced can be upgraded to a

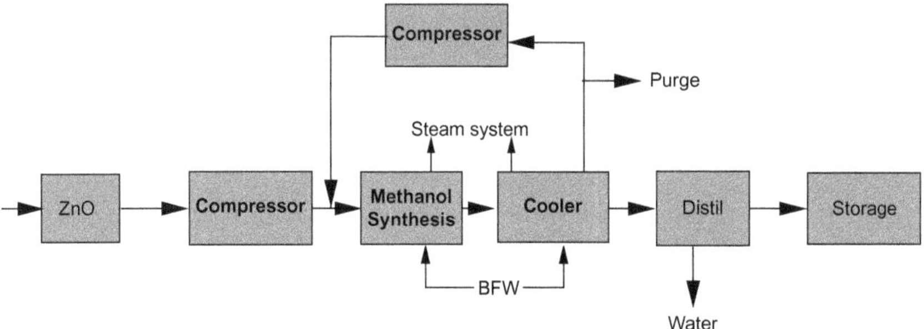

Figure 4.4. Layout for Methanol Synthesis

---

[9]The main manufacturing process uses steam reforming of hydrocarbons, particularly natural gas — steam methane reforming. If the feedstock is of a sustainable origin, for example, glycerol produced as a biodiesel byproduct, then this also will produce green methanol. See J.G. van Bennekom, R.H. Venderbosch and H.J. Heeres, *Biodiesel — Feedstocks, Production, Applications* (Z. Feng ed.) ISBN 978-953-51=0910-5.

chemical-grade product by distillation. The intermediate methanol is passed into storage. The reaction stoichiometry is

$$CO + 2H_2 = CH_3OH$$

and

$$CO_2 + 3H_2 = CH_3OH + H_2O.$$

To a good approximation, the quantity of water in the raw methanol is given by the second equation relating the water content to the concentration of carbon dioxide in the SYNGAS.

Methanol to Olefins[10]

The conversion of methanol into olefins[11] is similar to the commercially proven methanol to gasoline (MTG). This was commercialised using natural gas as the feedstock in New Zealand. The variant, which has been commercially proven in China, generally uses similar catalysts to produce only light olefins rather than the iso-paraffins and aromatics of the MTG process. This leads to the prospect of biomass conversion into green resins (solids). These high-value products may be easier to transport and sell than liquid fuels; Figure 4.5 illustrates the basic unit operations for the process.

Methanol, which need not be the highest-grade chemical methanol, is produced and stored before feeding to the methanol to olefins (MTO)

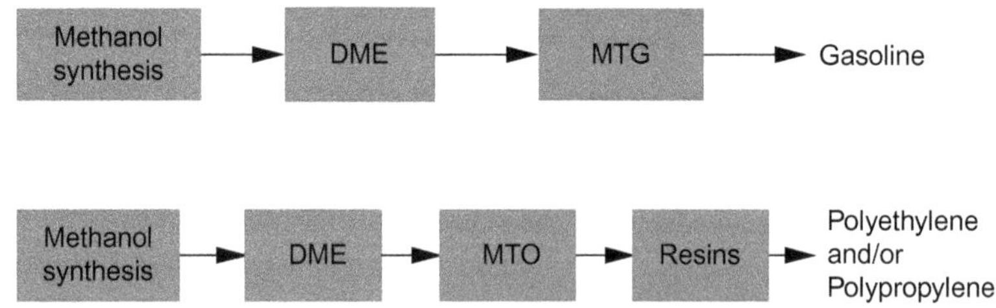

Figure 4.5. Process Steps for Converting Methanol into Gasoline and Olefins and Resins

---

[10] D. Seddon, *Petrochemical Economics — Technology Selection in a Carbon Constrained World*, ICP, 2010.

[11] M. Sutton and P. Roberts, *Hydrocarbon Processing*, 89 (July, 2007).

plant. The conversion of methanol into olefins is highly exothermic, and to help control heat evolution, some processes use a primary reactor to convert some of the methanol into dimethyl ether (DME) by the reaction:

$$2CH_3OH = CH_3.O.CH_3 + H_2O.$$

The MTO route can be optimised to produce either ethylene and propylene or solely propylene, for which there is strong and increasing demand. The basic stoichiometry for ethylene is

$$2CH_3OH = C_2H_4 + 2H_2O.$$

Higher olefins are produced by the reaction of ethylene with methanol:

$$CH_3OH + C_2H_4 = C_3H_6 + H_2O.$$

However, in detail, the conversion of methanol into olefins is quite complex.

Early Mobil Methanol to Olefins Processes

Early attempts to convert methanol into olefins were based on the zeolite ZSM-5. The Mobil MTO process was based on the fluidised bed version of the MTG technology. The conversion took place at about 500°C, allegedly producing almost complete methanol conversion. However, careful reading of the patent literature indicates that the complete methanol conversion may not have been achieved by this means. Because of incomplete conversion, there would be a necessity to strip methanol and DME from water and hydrocarbon products to recycle unconverted methanol. In this variant, the total olefin yield is less than 20% of the products, of which ethylene is a minor but not insignificant product. The major product is gasoline. Ethylene is difficult to process and had to be treated specially. Claims that it is possible that ethylene can be recycled to extinction conflict with the known behaviour of ethylene in zeolite catalyst systems and have to be viewed with some suspicion.

The methanol to chemicals (MTC) process was primarily designed to produce ethylene by operating an MTG-type catalyst and process at low pass conversion in a fixed bed reactor. The route was developed by the A.E.C.I. in South Africa, which demonstrated the process to pilot plant scale.

The principal reaction is brought about at low conversion in a series of reactors (10% conversion per reactor with ca. 40% conversion overall). The products, both aqueous and hydrocarbon phases, are heavily laden with methanol and DME, and as a consequence, extensive extraction and recycling were required.

The principal product is ethylene. The higher products are rich in olefins (66% C_3 + C_4 olefins, which are 41% of the total product). Like the Mobil MTO process, this process also produces a good-quality gasoline and a heavy gasoline, which may require hydrotreatment before use.

UOP Methanol to Olefins Process

The UOP process, developed jointly with Norsk Hydro/Statoil,[12] has been developed to a semi-commercial scale in Norway. The process uses proprietary catalysts based on a SAPO molecular sieve.

Two variants of the process are available, one maximising ethylene and the other propylene. The performance appears to be similar to that of the conversion of MTO using small pore zeolites. Such systems suffer from high methane yield (which has to be recycled back to a gasifier or reformer) and high coke yields. The formation of olefins is promoted by using crude methanol, which can contain up to about 17% water.

The coke formation leads to catalyst fouling. This is solved in the UOP process by continuously removing a portion of the catalyst and passing this to a separate regenerator. After regeneration by combustion of the coke in air, the catalyst is sent back to the main reactor. In concept, this is similar to fluid catalytic cracking of refinery stocks. The process layout is illustrated in Figure 4.6.

After separation of the mixed olefins, the product workup is similar to that in a steam cracker using liquefied petroleum gas (LPG) feedstock. Small amounts of carbon dioxide are removed, and the hydrocarbon gases

---

[12] M. Sutton and P. Roberts, "Petrochemical Processes 99," *Hydrocarbon Proc.* 125 (March, 1999).

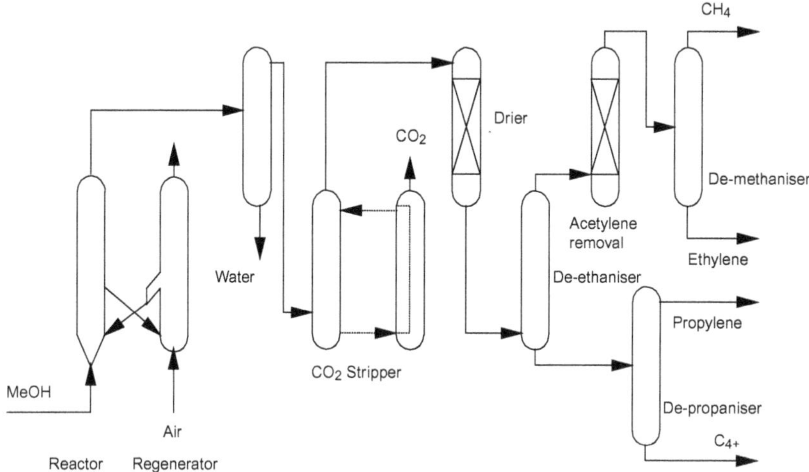

Figure 4.6.   UOP MTO Process Scheme

are dried before passing to a de-ethaniser column. The C_{2-} fraction is passed to an acetylene removal unit before methane is removed from the C_2 stream. This comprises 98+% ethylene, the remainder being ethane. The C_{3+} stream is split between the C_3 fraction (98% propylene) and C_{4+}. The workup of the C_4 stream to produce linear butenes (not shown in the figure) is likely to be less problematic than the corresponding C_4 stream from steam crackers, which is a highly complex mixture and cannot be separated by fractionation alone. The process produces little product above C_5.

The Lurgi Methanol to Propylene Process

The process has been demonstrated on a pilot scale by Lurgi and Statoil. Sufficient propylene has been produced to make polypropylene resin products by Borealis.[13] This process appears to use an oxide-doped ZSM-5 zeolite catalyst in fixed bed reactors. The oxide doping promotes the methanol conversion to olefins. All olefins other than propylene are

---

[13] Borealis A/S, *First Polypropylene Product Made from Natural Gas,* Press Release, September 16, 2003.

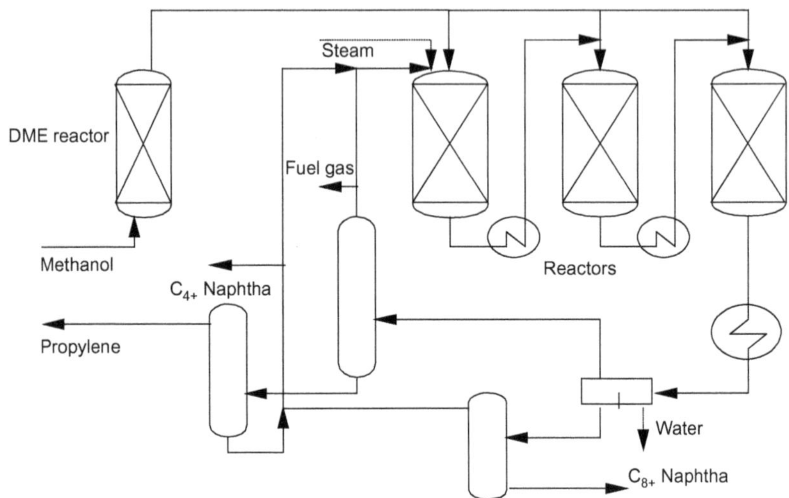

Figure 4.7. Lurgi MTP Process

recycled to extinction or purged as fuel gas or produced as naphtha. The flow sheet is illustrated in Figure 4.7.

Because fixed bed reactors are used, the heat of reaction must be removed. This is achieved by first converting some of the methanol to DME in the first reactor (similar to MTG) and then splitting the feed to a series of reactors. Overall, the method resembles the operation of a methanol quench converter where fresh feed is introduced at different points within a single reactor. The operation is at about 500°C, at which temperature propylene is favoured over ethylene. Overall promotion of olefin yield is obtained by adding steam. Downstream of the reactors is separation columns, which separate the C$_3$ product (ca. 80% propylene) from naphtha and fuel gases.

Comparison of Alternative Routes

A comparison of the yield of olefins from the routes discussed earlier is shown in Table 4.2.

The MTG process is aimed at producing gasoline (C$_5$–160°C cut). In the process, the light olefins are generally recycled or extracted as LPG.

Table 4.2. Comparison of Olefin-Producing Processes

	MTG	MTC	UOP MTO	UOP MTO	MTP	HT-FT
Ethylene	3.2	25.2	45.6	33.6	0.0	4.0
Propylene	4.7	16.5	29.6	44.6	67.9	11.4
Butenes	8.3	5.0	9.5	12.8	0.0	9.3
Total olefins	16.2	46.7	84.7	91.0	67.9	24.7
Fuel gas	21.2	15.6	5.6	2.0	6.1	17.8
C_5–160°C	58.0	33.0	5.5	5.5	26.0	32.6
160°C–350°C	5.0	1.0				13.0
>350°C						5.4
Water phase or coke		3.7	4.2	1.5	0.0	6.5

The MTC process produces high yields of light olefins and aromatic naphtha, which can be worked up to extract aromatics. This product slate is very similar to that for naphtha steam cracking. The product slate for the UOP/Statoil MTO process can be swung between ethylene-rich and propylene-rich products. Liquid products are much reduced, but there is carbon loss to coke. In the Lurgi methanol to propylene (MTP) process, all of the light olefins are recycled to extinction, but this increases the amount of fuel gas and naphtha product. For comparison, olefins produced by the high-temperature Fischer–Tropsch (HT-FT) process are included.

Conversion of Olefins into SAF

The oligomerisation of light olefins (propylene and butene) into gasoline, kerosene (jet) and diesel boiling range products is a commercial refinery process. Using renewably produced olefins would lead to green gasoline, green diesel and SAF. There are several alternative processes, the oldest being the oligomerisation brought about by acid catalysts, particularly phosphoric acid adsorbed on silica.[14] This process is usually used to produce polymer gasoline but can be adapted to produce higher boiling

---

[14] G. Egloff, "Polymer Gasoline," *Ind. Eng. Chem.* **28**(12), 1461–1467 (1936).

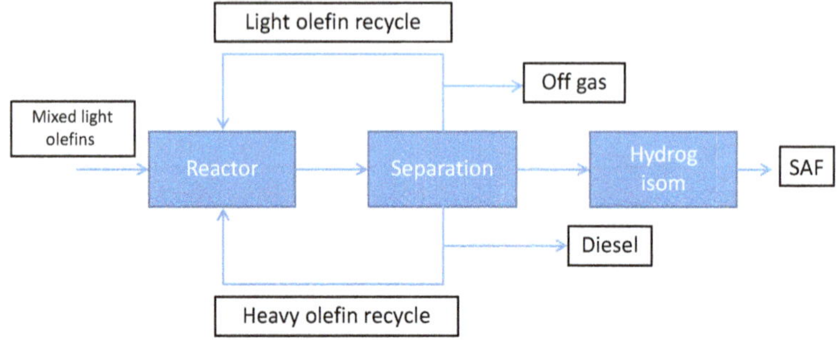

Figure 4.8. SAF by ExxonMobil Process

products, such as jet fuel and diesel. The latter process was used in South Africa to produce jet and diesel fuel from the light olefins products produced by the HT-FT process (Table 4.2).[15] The polymerisation of olefins generates a branched olefinic product. For optimum SAF properties, this should be hydrogenated to remove the olefinic bond.

Another approach to the production of SAF using olefins is the Mobil olefins to gasoline and distillate (MOGD) process,[16] now proposed by ExxonMobil. This has been adapted to focus more on the production of the jet fuel cut;[17] the scheme is outlined in Figure 4.8.

Mixed light olefins (C_2–C_5), produced in a sustainable manner, enter the reactor system and are oligomerised to higher olefins boiling in the gasoline to diesel range. This is passed to a separation (distillation) section. Light olefins are recycled to the reactor with a purge to remove cracked gases. The jet cut is hydro-isomerised to produce the SAF. Heavy olefins are also recycled but are slow to react, so part of this stream is drawn off to produce green diesel.

The main issue with this approach is the production of the mixed olefin feedstock. Ethylene is slow to react to the preferred olefins, which

---

[15] M.E. Dry, "The Fischer–Tropsch Synthesis," in *Catalysis Science and Technology* (J.R. Anderson and M. Boudart eds.) Vol. 1, Springer-Verlag, 1981.

[16] W.E. Garwood and W. Lee, *US Patent* 4,227,992 October 14, 1980 (to Mobil Oil Corporation).

[17] J.H. Beech Jr, H. Owen, M.P. Ramage and S.A. Tabak, *US Patent* 5,073,351, Dec. 17, 1991 (to Mobil Oil Corporation).

are propylene and higher olefins. This mixture can be produced using the Mobil MTO process, with methanol being green methanol produced by gasification as discussed earlier or steam reforming of glycerol (a by-product of biofuel production) or steam reforming of other light hydrocarbons produced from biomass. In theory, ethanol could be used, but this generally reacts slower than methanol and has not been demonstrated.

SAF Economics from Methanol

The economics of the conversion of ethylene into SAF has been discussed earlier, so the approach to the use of methanol as a feedstock is what the cost is for the production of olefins from methanol. First, there are some remarks about the cost of production of methanol from biomass.

Economics of Methanol Production

As of writing, most methanol is made in countries with access to large quantities of low-cost natural gas by the SMR process. Low-cost gas production underlies the low traded price of methanol in the range of $250 to an average in the region of $350/t (Figure 4.9).

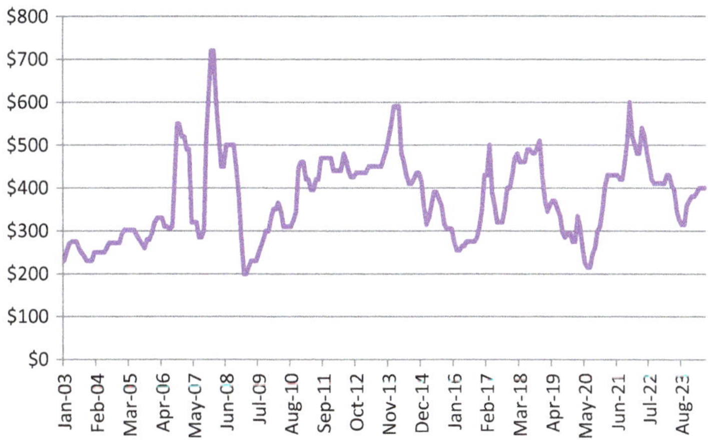

Figure 4.9. Methanex Posted Asian Contract Price for Methanol ($/t)

An analysis of biomass economics[18] is given later for the production of synthetic crude oil. This has been adapted for the conversion into methanol. The summary economics for production from natural gas and biomass is given in Table 4.3.

For natural gas, the process used is SMR to produce the SYNGAS. The manufacturing cost using partial oxidation is similar in cost to that for SMR. The price of the feedstock gas is taken as $2/GJ, which is typical of gas costs for large-scale export-orientated methanol plants producing 850 kt/y or more. The methanol production cost is estimated at $285/t.

For biomass, feedstock supply and transport logistics limit available feed to about 1 Mt/y, and this can be used to deliver a methanol production of 587 kt/y. This production would be a maximum and comprises two parts. The first is production from the raw gas from the gasifier of 437 kt/y, and the second component produced from methane in the raw SYNGAS of 150 kt/y. Methane can be processed separately by SMR or by recycling to the gasifier. Although producing less product than the natural gas route, the biomass gasifier and its ancillary process plant are considerably more expensive than the SMR. This is not compensated by the lower cost of the feedstock. The result is an estimate of $490/t for methanol produced by biomass.

Olefins from Methanol

The technology of olefin production from methanol has been previously reviewed and estimated.[19] The data has been updated and brought to a 2022 cost basis. The sensitivity of the production costs for olefins as a mixture of ethylene and higher olefins is illustrated in Figure 4.10. The cost for the production of propylene only is expected to be similar or slightly lower.

For methanol prices generated by low-cost natural gas, the cost of olefin production is expected to be in the range of $1,150 to $1,300/t. For the higher cost of methanol produced by biomass gasification, the cost is expected to be about $1,900/t.

---

[18] T.G. Kreutz, E.D. Larson, G. Lui and R.H. Williams, *Fischer–Tropsch from Coal and Biomass,* 25th Annual Pittsburgh Coal Conference, 2008.

[19] D. Seddon, *Gas Usage and Value, the Technology and Economics of Natural Gas Use in the Process Industries*, PennWell, 2006 and *Petrochemical Economics — Technology Selection in a Carbon Constrained World*, Imperial College Press, 2010.

Table 4.3. Estimate of the Production Cost of Methanol from Natural Gas and Biomass

		Natural gas	**Biomass**
Nominal capacity	kt/a	850	587.32
	kt/d	2.50	1.73
	PJ/a	19.30	13.33
	bbl/d		
Construction period	y	3	3
Required return	%	10.00%	10.00%
Operating period	y	20	20
Capital recovery	%	14.60%	14.60%
Base capital cost	$M	250	675.92
Base year		1990	2007
Location factor		1	1
Index	2022	3.1740	1.8467
Final CAPEX	$M	793.50	1248.24
Return on capital	$M/a	115.85	182.24
Unit WC charge	$/t	300.00	300.00
Working capital (% Capex or 30 days)	$M	22.50	14.71
Return of working capital	$M/a	2.25	1.47
Capital charges (C)	$M/a	118.10	183.71
Operating Costs	$M/a		
Labour	3% Capex/y	23.80	37.45
Maintenance	3% Capex/y	23.80	37.45
Insurance	1.5% Capex/y	11.90	18.72
Catalysts and chemicals	$M/a	0.25	0.25
Non-feed OPEX (O)		59.76	93.87
Total non-feed costs	$M/a	177.86	277.58
Gas usage	PJ/a	32.16	19.403
Efficiency	%	60.00%	68.71%
Feedstock price	$/GJ	2.00	0.53
Feedstock costs	$M/a	64.32	10.35
Production cost calculation			
Gross costs	$M/a	242.18	288.01
By-product credits (B)	$M/a	0	0
Net costs	$M/a	242.18	288.01
Unit production cost	$/t	284.92	490.39

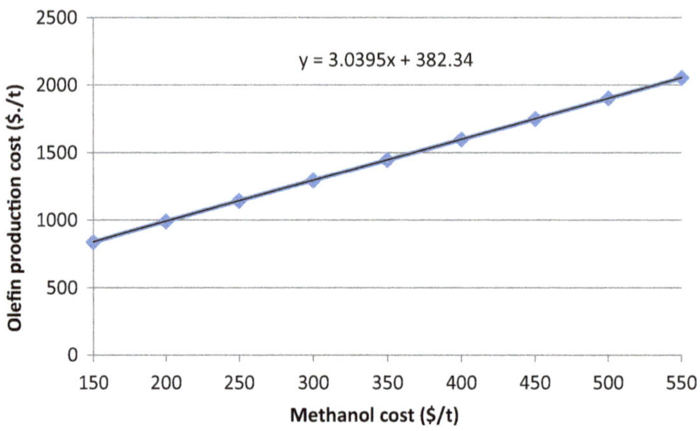

Figure 4.10.   Sensitivity of Olefins Production Cost to Methanol Cost

Production Costs of SAF from Olefins Produced from Methanol

Several processes for the production of SAF from olefins produced via the methanol route have been discussed earlier. The economic analysis presented here is based on the older polymerisation route to gasoline and kerosene.[20] It is assumed that this process can be modified to maximise the kerosene fraction by recycling the gasoline fraction. This will then be hydrogenated to remove olefinic bonds in the product. The preferred feedstock is C3 and C4 olefins rather than ethylene. A possible flow sheet is illustrated in Figure 4.11.

Fresh olefin feed enters the oligomerisation reactor and mixed with the recycle gasoline fraction and any unconverted light olefins. There may be several rectors in series for maximum conversion. The products are distilled, and the gasoline and unconverted olefins are recycled. Any cracked gas (expected to be minimal) will be discharged from the top of the column. The column bottoms, distilling in the kerosene boiling range, are mixed with hydrogen and passed to a hydrogenation unit, removing the olefinic bonds from the product and producing SAF. Excess hydrogen and any cracked products are discharged from the top of the column.

---

[20] Axens Oligomerization — Polynaphtha Process in *Hydrocarbon Processing*, Refining Processes 2002, November 2002, p. 140.

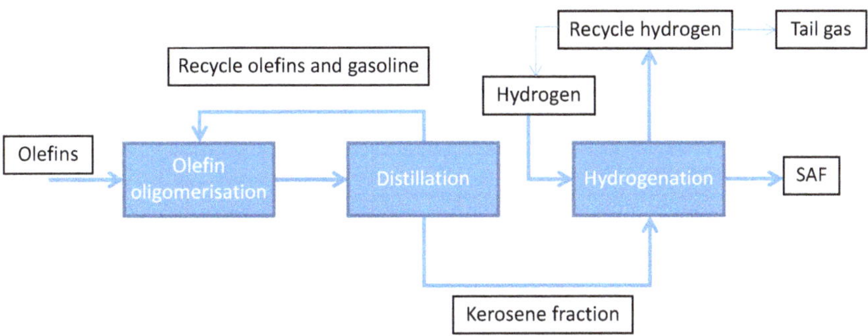

Figure 4.11. Possible Flow Sheet for Oligomerising Olefins to SAF

The data has been updated to 2022 values, and the statistics are given in Table 4.4.

All products are expected to be SAF. The facility has a nominal output of 5,000 bbl/d. The 2022 IBSL Capex is estimated to be $20.14 M; this value is based on a 2002 estimate for an oligomerisation plant-making polymer gasoline; it may be considered low since the full cost of the final hydrogenation step is not included, but this is expected to be a small addition. This ISBL cost is taken as 60% of the total capital cost of a greenfield facility. Since the product will be highly branched, it is assumed that an isomerisation step is not required.

The efficiency of the process to produce SAF is assumed to be 95%, thus requiring 210,000 t/y of olefins and about 2,400 t/y of hydrogen at $3,000/t. For the case study in Table 4.4, the olefin cost is $1,200/t, which is the cost from low-cost (i.e. natural gas or coal-based) methanol (Figure 4.10). This gives an estimate for the SAF production cost of $159/bbl.

Note the dominant cost is that for the olefin feedstock (94% of the total costs). The economics is not that sensitive to the capital cost of the facility or other factors. The sensitivity to the cost of the olefin is illustrated in Figure 4.12.

As estimated earlier, methanol produced by the gasification of biomass will be nearly $500/t, which will result in olefin costs nearly the $2,000/t mark. Olefins at this price point will result in an SAF production cost of about $250/bbl, which is more than double the cost of jet fuel from crude oil at $100/bbl.

Table 4.4. Statistics for Production of SAF via Olefin Oligomerisation

SAF scale	bbl/d	5,000
Density	kg/L	0.74
SAF	kg/d	588,226
SAF	t/y	199,996
CAPEX ISBL 2002	$M	8.5
2002 index		2.369
2022 Capex	$M	20.14
Greenfield Capex	$M	33.56
ROC % Capex (2, 10, 20)	% Capex	13.84%
ROC	$M/y	4.64
Working capital (30d, $100/bbl)	$M	15.00
Return on WC (10%)	$M/y	1.5
Olefin required ideal	kt/y	200.00
Efficiency		95%
Olefin actual	kt/y	210.52
Olefin price	$/t	1,200
Olefin cost	$M/y	252.63
Hydrogen ideal	kt/y	2.38
Hydrogen efficiency		90%
Hydrogen actual	kt/y	2.65
Hydrogen price	$/t	3,000
Hydrogen cost	$M/y	7.94
Operating costs %Capex		
Labour $M/y	2%	0.67
Maintenance $M/y	3%	1.01
Insurance etc. $M/y	1.50%	0.50
Catalysts and chemicals $M/y	1%	0.34
OPEX	$M/y	2.52
Total costs	$M/y	269.23
	$/bbl	158.37

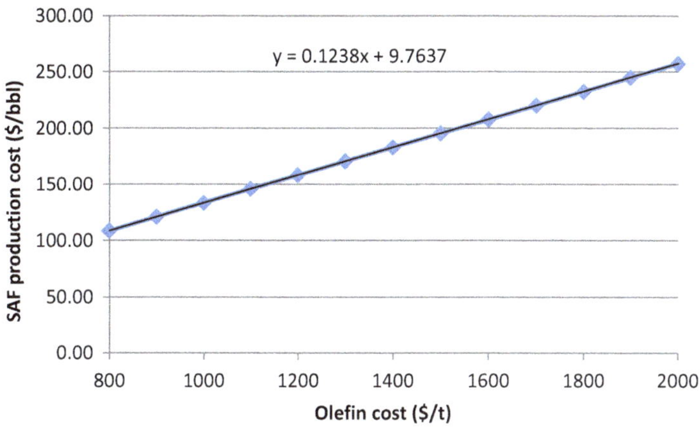

Figure 4.12. Sensitivity of SAF Production Cost to Olefins Cost

Fischer–Tropsch Process for Producing SAF

SYNGAS produced from biomass can be converted into synthetic hydrocarbon fuels, including SAF using the FTP. The literature on this process is very extensive, and only the basic technology targeted at producing SAF will be discussed.[21]

The FTP concerns the conversion of SYNGAS into hydrocarbons and water over promoted transition metal catalysts, particularly iron and cobalt. The basic reaction is

$$n\mathrm{CO} + 2n\mathrm{H}_2 = -[\mathrm{CH}_2]_n- + n\mathrm{H}_2\mathrm{O},$$

where $-[\mathrm{CH}_2]_n-$ represents a hydrocarbon chain with n ranging from 1 to typically 35. At its basics, the reaction is a polymerisation reaction that involves the insertion of a CO moiety into a growing carbon chain attached to the catalyst. The $-\mathrm{CO}-$ group is reduced to $-\mathrm{CH}_2-$. The process either continues or the chain breaks away from the catalyst. The term ALPHA is the ratio of the probability of the chain growing versus the chain breaking away.

ALPHA is the ratio of molecules with carbon number n divided by the moles of carbon number $n - 1$. ALPHA is less than 1. For SAF, we require

---

[21] M.E. Dry, "The Fischer–Tropsch Synthesis," in *Catalysis Science and Technology* (J.R. Anderson and M. Boudart eds.) Vol. 1, Springer-Verlag, 1981 and references therein.

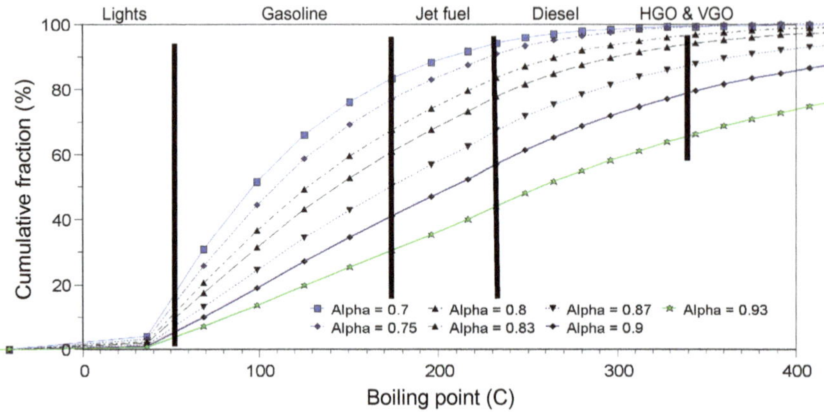

Figure 4.13. Influence of Alpha Value on the Volume Fraction of Various Refinery Cuts

products in the range $n = 8$ to 15. In general, the product slate from a Fischer–Tropsch synthesis resembles a synthetic crude oil. Figure 4.13 illustrates how the change in alpha value influences the volume of various fractions.

The alpha value is influenced by various parameters of the synthesis — catalyst, pressure, temperature, and so on. Temperature is the main parameter; high temperatures favour low alpha values and low temperatures favour high alpha values.

With a low alpha value (0.70), most of the liquid product lies in the gasoline range, whereas a high alpha value (>0.90) leads to large volumes boiling in the diesel range and higher (wax) ranges. From the figure, note that for the intermediate ranges of alpha (0.8–0.9), the volume fraction of jet range product is more or less constant at about 15% to 20% of the total fraction. This is referred to as the straight cut fraction and is clearly inadequate for producing large volumes of SAF.

There are two ways to increase the volume of the jet-fuel cut:

1. Operate at high temperature resulting in low alpha value producing an excess of the gasoline (naphtha cut) together with some light products (C_3 and C_4). High-temperature operation will also result in these fractions being rich in olefins. From the separated light, gas and naphtha cut covert any paraffins to olefins and oligomerise the olefins into the

jet fuel range. Add this to the straight-run jet produced. This will also produce a small amount of diesel and practically no wax.

2. Operate at low temperature, preferably producing a product of very high alpha value (>0.90). This product is dominated by the high boiling wax fractions. The fractions are not necessarily separated; rather, the entire synthetic crude oil is passed to a hydrocracker with the duty to split the heavier wax chains into the diesel and jet fuel fractions.

These approaches follow the experience for maximising the diesel cut from the current commercial Fischer–Tropsch operations, which is shown figuratively in Figure 4.14.

FTPs operating at an intermediate alpha value, as indicated in Figure 4.13, produce limited quantities of jet and diesel.

The Synthol HT-FT synthesis developed by Sasol in South Africa, operating at an alpha value of about 0.7, produces an olefin-rich gas and naphtha as its main products. Paraffins are separated and dehydrogenated to olefins, and the combined olefin fraction is oligomerised into kerosene (jet) and diesel.

The Shell MDS process operates at low temperature, delivers a product with a very high alpha value (possibly >0.95), and produces a wax product. This is not separated but is hydrocracked into mainly diesel. The Sasol low-temperature slurry-bed process produces a similar heavy wax product.

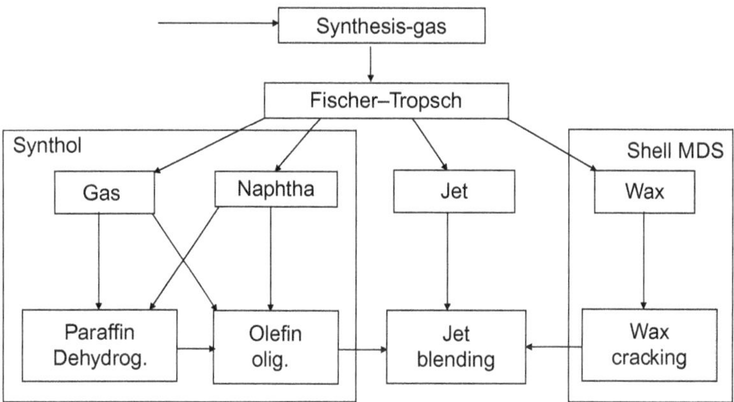

Figure 4.14. Approach to Maximising the Jet Cut from the Fischer–Tropsch Process

The optimum route for producing diesel from the standpoint of carbon efficiency is the low-temperature synthesis followed by the wax-cracking route. However, this may not be the optimum for jet fuel production, as it would imply that a softer (lower boiling point) wax should be produced to increase the yield of the lower boiling (relative to diesel) range jet-fuel cut when cracked; the production of lower boiling wax would inevitably result in a rise in the production of light gases in the Fischer–Tropsch synthesis.

Economics of Biomass to SAF via Fischer–Tropsch Process

The conversion of biomass to transport fuels by the FTP has been extensively studied. These studies lay out the main issues for the conversion of biomass into hydrogen and carbon monoxide. The production of liquid products by the FTP requires biomass gasification to SYNGAS, WGS, followed by the Fischer–Tropsch conversion to liquids. Many of these studies also address the issue of biomass collection and logistics and hence limit to the size of the facility.

The following economic descriptive has been adapted from a study by Kreutz *et al.*,[22] for the conversion of 1 Mt/y of corn stover into Fischer–Tropsch liquids. A proximate and ultimate analysis for the corn stover is given in Table 4.5. On a dry basis, the stover had a specific energy of 18.7 GJ/t (higher heating value [HHV]).

Feedstock and product flow rates are given in Table 4.6. The synthetic crude production (diesel plus gasoline) is given as 4,415 bbl/d. The ideal production rate is approximately 5,475 bbl/d, which indicates that there could be some value in process optimisation.

A power balance for the system is given in Table 4.7. The steam and power system is capable of producing about 66 MW of power, of which 27 MW is used within the facility. The use is dominated by the power demand of the ASU and the attached oxygen compressor (total 21 MW). The SYNGAS clean-up (Rectisol™) consumes about 4 MW.

There is an excess of power generation over the demand of nearly 49 MW, which is exported from the facility.

---

[22]T.G. Kreutz, E.D. Larson, G. Lui and R.H. Williams, 25th Annual Pittsburgh Coal Conference, 2008.

Table 4.5. Proximate and Ultimate Analysis of the Corn Stover Feedstock

Corn stover	Mt/y	1
<u>Proximate</u>	As received	
Fixed C	wt%	18.10%
Volatile matter	wt%	61.60%
Ash	wt%	5.30%
Moisture	wt%	15.00%
Total		100.00%
LHV	MJ/kg	14.509
HHV	MJ/kg	15.935
<u>Ultimate</u>	dry basis	
Carbon	wt%	46.96%
Hydrogen	wt%	5.72%
Oxygen	wt%	40.18%
Nitrogen	wt%	0.86%
Sulphur	wt%	0.09%
Ash	wt%	6.19%
Total		100.00%
HHV	MJ/kg	18.748

Table 4.6. Feedstock and Product Flow Rates

Stover as received	t/d	3,581
Dry	t/d	3,044
DAF	t/d	2,822
	MW LHV	601
	MW HHV	660
Oxygen	t/d	59
Purity	mass%	99.54%
Oxygen for gasifier	t/d	58
Diesel	bbl/d	2,546
Gasoline	bbl/d	1,869
Total product	bbl/d	4,425

Table 4.7. Power Balance for Biomass to Hydrogen Facility

Production	MW	65.99
Demand	—	—
Biomass	MW	0.41
Lock hopper	MW	0.35
Synthesis gas clean-up	MW	3.86
Fuel gas comp	MW	0.08
Steam cycle	MW	1.27
ASU	MW	14.78
O_2 Compressor	MW	6.55
Total consumption	MW	27.3

Table 4.8. Capital Cost Breakdown for a Biomass to Synthetic Crude Facility

		2007		2022
Land	$M	5.00	0.58%	$9.23
Gasifier	$M	266.00	30.61%	$491.23
Rectisol and WGS	$M	58.00	6.67%	$107.11
ASU	$M	94.00	10.82%	$173.59
Rectisol	$M	152.00	17.49%	$280.70
Power Island	$M	64.00	7.63%	$118.19
		639.00	73.53%	$1,180.06
Engineering	+15%	95.85	11.03%	$177.01
Contingency	+10%	63.90	7.35%	$118.01
Start up, spares	+6%	38.34	4.41%	$70.80
Off-sites	+5%	31.95	3.68%	$59.00
TOTAL	$M	869.04	100%	$1,604.88

The capital cost breakdown is given in Table 4.8. The cost estimates for 2007 have been escalated to 2022 values in the final column. The capital cost estimate is of $1,605 M for 2022 construction.

Using the above statistics for a biomass to the synthetic crude (as a mixture of gasoline and diesel), the production cost is developed in Table 4.9. It is assumed that the facility will take three years to build and

Table 4.9. Estimate for the Cost of Synthetic Crude from Biomass

Capacity	bbl/d	4,415
2018 CAPEX	$M	$1,604.88
ROC (20 y, 10% DCF 3 y build)	%	14.60%
ROC (20 y, 10% DCF 3 y build)	$M/y	$234.35
Working Cap (30d; 10%)	$M	$13.25
	$M/y	$1.32
OPEX	% Capex	
Labour	1%	$16.05
Maintenance	3.00%	$48.15
Insurance	1.50%	$24.07
Catalysts and chemicals	1%	$16.05
	$M/y	$104.32
Fixed costs	$M/y	$339.99
Feedstock	Mt/y	$1.00
Feedstock cost	$/t	$10.00
Feedstock costs	M$/y	$10.00
Total annual costs	$M/y	$349.99
By-product electricity	MW	38.69
	MWh/y	315,710.4
	$/MWh	50
	$M/y	$15.79
Net cost	$M/y	$334.21
Production cost	$/bbl	$222.64

require a 10% discounted cash flow return, operate for 20 years and pay a 2% capitalised royalty that will require a return on capital (ROC) of 14.6%. Working capital is estimated for 30 days of product storage with a product value of $100/bbl and requires a return of 10%/y.

Non-feedstock operating costs are $104 million/year. The feedstock (1 Mt/y) is assumed to cost $10/t to cover collection and delivery costs. The total annual costs of $350 million are reduced by $15.79 million/year from the sale of by-product electricity at \$50/MWh. The net production costs are \$334 million/year. This generates a synthetic crude oil (gasoline, jet fuel and diesel) production cost of \$222/bbl.

Production of SAF from Carbon Dioxide

There is an increasing interest (mainly academic) in the production of SAF using carbon dioxide as the source material for carbon. This can be derived from

- Waste industrial streams, such as from the production of ammonia from natural gas or coal. Carbon dioxide is widely produced in this manner for use in enhanced oil recovery operations and is commercially proven.
- Work is in progress to extend the capture of carbon dioxide from flue gases and similar gases of low carbon dioxide concentration. This is considerably more costly than capture from industrial waste streams, where carbon dioxide concentration can be above 90%. This field is yet to be commercially proven.
- Total combustion of biomass in air or oxygen followed by carbon dioxide capture. Again, this is as yet to be commercially proven.
- The extraction of carbon dioxide from air. The main issue is the very low concentration of the carbon dioxide (0.04%). This requires windy locations, which helps transport the large volumes of air through the extraction plant and a cheap source of renewable energy for regenerating absorbents.[23]

Carbon dioxide is reacted with hydrogen produced by renewable means[24] to produce either methanol or hydrocarbons via the FTP.

For methanol, as noted in the aforementioned section, the conversion of methanol on the conventional $Cu/Zn/Al_2O_3$ catalyst appears to proceed by hydrogenation of carbon dioxide; however, carbon dioxide in the feed over about 20% leads to catalyst oxidation and inhibition. Therefore, over

---

[23] Climeworks, (https://climeworks.com) have a commercial operation in Iceland sequestrating 4,00t/y carbon dioxide.

[24] D. Seddon, *The Hydrogen Economy*, World Scientific, 2022.

such catalysts, it will be important to reduce the carbon dioxide concentration by first applying the reverse WGS reaction to generate a SYNGAS high in carbon monoxide.

$$CO_2 + H_2 = CO + H_2O.$$

Once produced, the methanol can be used to produce SAF in the methods discussed earlier.

Over the many decades of the synthetic methanol industry, there has been extensive research into catalysts for methanol synthesis. Some of the past attempts may be more applicable to a high carbon dioxide SYNGAS;[25,26] an analysis of this subject is beyond the scope of this book.

To produce hydrocarbons using the FTP, it will be necessary to go further and eliminate the carbon dioxide from the feed gas. Not only does carbon dioxide oxidise and inhibit the catalyst, but also high levels of carbon dioxide tend to lead to methanation rather than the production of hydrocarbons.

Once produced, the SYNGAS of the optimum stoichiometry for downstream operations is fed to a Fischer–Tropsch hydrocarbon facility to produce SAF in the manner described earlier.

Again, for many decades, since the discovery of the FTP, there has been extensive research into catalysts, some of which may be more applicable to the production of hydrocarbons from carbon dioxide than current iron- or cobalt-based formulations.[27,28]

The Production of High-Density Kerosene

All of the processes discussed earlier produce a purely paraffinic product. This may be acceptable to some users, but the density of the product is too low compared to that required for an ideal jet fuel complying with the

---

[25] G.C. Chinchen, P.J. Denny, J.R. Jennings, M.S. Spencer and K.C. Waugh, "Synthesis of Methanol: Part 1. Catalysts and Kinetics," *Appl. Catal.* **36**, 1–65 (1988).

[26] Noerma J. Azharia,1, Denanti Erikab,1, St Mardianab, Thalabul Ilmic, Melia L. Gunawana,c, I.G.B.N. Makertiharthaa,c, Grandprix T.M. Kadja, Methanol Synthesis form CO2: A mechanistic overview, *Results Eng.* **16**, 100711 (2022).

[27] R.B. Anderson, *The Fischer–Tropsch Process*, Academic Press, 1984.

[28] M.E. Dry, "The Fischer–Tropsch Process — Commercial Aspects," in *Catalysts Today* (Hutchings and Scurrell eds.), **6**(3), (January, 1990).

Jet-A standards. The low density of the fuel leads to a lower limit on the range of the aircraft. Although commercial jet fuel is mainly paraffinic, it contains some level of aromatics, which increases the density of the fuel. This limits the use of paraffinic products (e.g. Fischer–Tropsch) kerosene to about 50% of a fuel blend, the rest being conventional mineral jet fuel. There seem to be three approaches to the problem:

1. Increase the fuel storage of the aircraft. At the time of writing, to promote very long-range flights, some airlines are seeking to install additional fuel tanks in the body of the aircraft. This will obviously be at the expense of other storage. This plan could be adopted for low-density SAF.
2. Change the fuel standard to facilitate low-density SAF. This will obviously restrict flight range, but many airlines and routes could accommodate this.
3. Develop renewable routes to aromatic kerosene blendstock, which will deliver a high-density blendstock for low-density SAF. This will deliver a truly drop-in fuel.

High Aromatic Kerosene Blendstock

The challenge to produce a drop-in jet fuel is to source suitable aromatics from a renewable feedstock to blend with the paraffinic fuels produced by the above-discussed processes.

One solution that is coming to the fore derives from efforts to produce synthetic fuels from lignocellulosic materials sourced from pertinent biomass.[29] In this approach, a three-step approach is used as illustrated in Figure 4.15.

The three stages are

1. Hydrogenate the biomass to break up the large macromolecules of the biomass into smaller oxygen-containing molecules.
2. Hydrogenate further to remove some of the oxygen and produce mainly acids, esters and alcohols and some paraffins.

---

[29] B. Zhang and D. Seddon, *Hydroprocessing Catalysts and Processes — The Challenges for Biofuels Production*. World Scientific Publishing Europe Ltd., 2018.

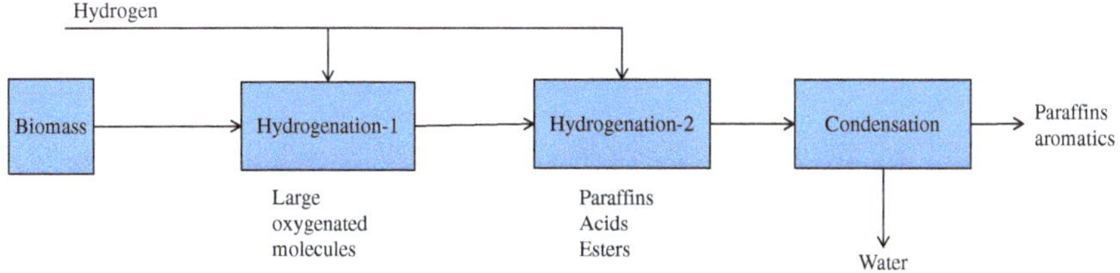

Figure 4.15. Conversion of Biomass to Aromatics

3. Use a condensation catalyst to remove all of the oxygen as water and produce a mixture of paraffins, olefins and aromatics.

Virent Inc. (a subsidiary of Marathon Petroleum Company) promotes a "BioForming" technology to accomplish this process. The process has several forms targeting either jet fuel and diesel products or chemicals such as *para*-xylene from sugar or biomass.[30]

The jet fuel blending product, known as synthesised aromatic kerosene (SAK), is blended with synthetic paraffinic kerosene, which contains no aromatics, at about the 20% level. This increases the resulting fuel's density to make it compliant with the Jet-A standards and produce fully renewable SAF.

Costs

At the time of writing, there is a paucity of data to assess the cost of producing an aromatic feedstock by this route, so an *a priori* method is used as follows.

Consider the conversion of sucrose into hexylbenzene. Sucrose is considered an ideal and widely available feedstock of known price, and hexylbenzene is a proxy for the type of compound we wish to produce; it has the correct properties for a jet-fuel component, namely:

- a high density of 0.861 kg/L
- an acceptable boiling point of 226°C

---

[30] See for example, P. Blommel and R. Cortright, *US Patent* 2014/03500317 and P.G. Blommel, B. Dally, W. Lymann and R.D Cortright, *US Patent* 11,130,914 (September 28, 2021) to Virent Inc. and references therein.

- an acceptable melting point of −61°C
- an acceptable flash point of 83°C.

We apply the above process to one mole of sucrose with the addition of 9 moles of hydrogen into one mole of hexylbenzene and 11 moles of water:

$$C_{12}H_{22}O_{11} + 9H_2 = C_6H_{13}.C_6H_5 + 11H_2O.$$

For this ideal situation, the cost of hexylbenzene production is illustrated in Table 4.10.

Table 4.10. Estimates for Production of Sustainable Aviation Kerosene

		Ideal	Eff.[a]	Expected	Eff.[a]	Corn syrup	Eff.[a]	Wood[b]
Inputs								
Sucrose	t	342	98%	348.9	73%	468.4	60%	570
Hydrogen	t	18	90%	20	85%	21.17	85%	21.17
Outputs								
C8+ products	t	162		162		162		162
Water	t	198		198		198		198
Sucrose	$/t	500		500		400		145
	$	171,000		174,489.8		187,397.3		82,650
Hydrogen	$/t	3,000		3,000		3,000		3,000
	$	54,000		60,000		63,529.41		63,529.41
Total feed inputs	$	225,000		234,489.8		250,926.7		146,179.4
Other costs (20% feed)	$			46,897.96		50,185.33		29,235.88
				281,387.8		301,112		175,415.3
C8+ products	$/t	1,388.8		1736.9		1,858.7		1,082.81
	$/kg	1.388		1.736		1.858		1.08281
	$/L	1.195		1.495		1.60		0.932
	$/bbl	190.0		237.6		254.2		148.13

[a]Carbon or hydrogen efficiency; [b]Wood pellets

The analysis is based on an ideal sucrose conversion as detailed in the ideal column. Sucrose (342t) and hydrogen (18t) produce C8+ product (hexylbenzene, 162t) and water (198t). With a sucrose price of $500/t, the sucrose feedstock costs $171,000. With hydrogen at $3,000/t, the hydrogen co-feed costs $54,000. The cost of production on an ideal feedstock-only basis for the C8+ product (hexylbenzene) is $1,388/t, which is $190/bbl, that is, the feedstock-only cost of the aromatic kerosene is higher than the typical traded price of jet fuel by about $50/bbl, with crude oil at $100/bbl.

A better estimate for the production cost will take into account the efficiencies of the process. For sucrose, which is a pure chemical feedstock, the efficiencies are likely to be high and are placed at 98% for the sucrose and 90% for the hydrogen (considering hydrogen's properties and handling under pressure will lead to higher losses). This leads to the higher usage rates given in the "expected" column. Some account should be taken of capital and non-feedstock operating costs. These capital and operating costs are unknown, but analysis of many chemical plant economics indicates that feedstock costs represent 80% of the production costs. Hence, an additional 20% of the feedstock costs is added to cover these segments. This generates a more realistic estimate for the production cost of $237/bbl for the aviation kerosene.

Following this case, another case is the use of corn syrup as described by Blommel et al.,[31] which is developed in the "corn syrup" column. The carbon efficiency is taken from the data provided, and the hydrogen efficiency is estimated. With corn syrup being 80% of the price of sugar, this produces an estimated cost of aviation kerosene production of $254/bbl.

To further reduce the production cost, we might consider the use of wood in the form of pellets or chips as the feedstock source. The ultimate analysis of wood pellets is similar to that for sucrose (wood carbon 46.8% c.f. sucrose 42.1%; wood hydrogen 5.61% c.f. sucrose 6.43%; wood oxygen 47.33% c.f. sucrose 51.46%).[32] Although similar to sucrose, the use

---

[31] P.G. Blommel, B. Dally, W. Lyman, R.D. Cortright, *US Patent* 11,130,914 (September 28, 2021) to Virent Inc., Examples 19 & 20.

[32] H. Sutanto, S. Suyitno, W.E. Juwana and T.G. Nurrohim, "Design, Production Cost, and Air Flow Distribution of Biomass Pellet Furnace," *Mekanika: Majalah Ilmiah Mekanika* **20**(2), 68–76 (2021).

of an inferior feed (containing impurities, ash and moisture) such as wood would be expected to result in a higher loss, the carbon efficiency has been reduced to 60% for this case. Wood pellets are priced in the region of about \$150/t. These lead to an estimated aromatic kerosene price of about \$150/bbl, which is similar to the price of aviation kerosene.

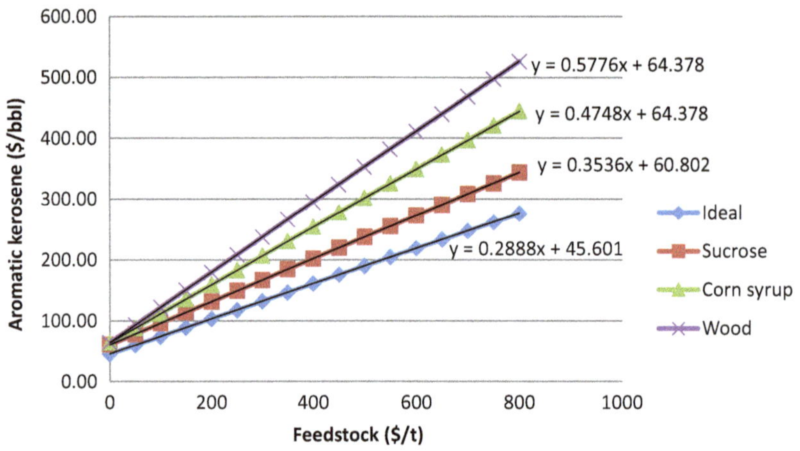

Figure 4.16.	Sensitivity of Aromatic Kerosene Production Cost to Feedstock Price

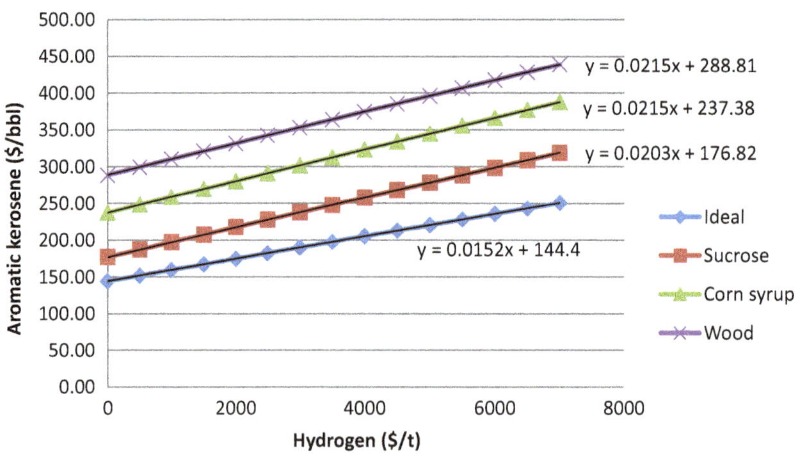

Figure 4.17.	Sensitivity of Aromatic Kerosene to Hydrogen Costs

The sensitivity of these cases to the cost of the feedstock is illustrated in Figure 4.16.

The slope of the sensitivity curves is a function of the process efficiency; low efficiency gives a greater slope. The use of lower-efficiency feedstocks, such as wood pellets, is dependent on the low price for the feedstock.

The sensitivity to hydrogen price is illustrated in Figure 4.17.

The importance of the sensitivity to hydrogen is that a requirement for many renewable energy projects is that renewable hydrogen be used. For most cases, renewable hydrogen will cost more than the $3,000/t used in the case study. If the hydrogen costs double ($6,000/t), this will considerably increase the production cost of the aromatic kerosene.

5

SAF BY PYROLYSIS OF BIOMASS

Pyrolysis processes heat the biomass in the absence of oxygen to produce a gaseous product, liquid products and a solid product referred to as biochar. The temperature of the pyrolysis ranges from 200°C to 700°C. If the temperature is on the lower end of the range of 200°C to 350°C, the process is slower. This process is generally referred to as torrefaction.

Pyrolysis is often classified into various types depending on the time residence in the reactor. The products from various forms of pyrolysis are generally similar in character, with the proportions of the different products varying according to the process parameters. Table 5.1 illustrates various types of pyrolysis.

Several versions of each type are under development, with many focusing on the production of biochar. This has several uses in the agricultural industries. For SAF and biofuels, we are mainly interested in the oil fraction. The pyrolysis route is attractive because the liquid products have a high level of aromatics present, which can be worked up to produce a high-density aromatic kerosene useful for blending with low-density paraffinic kerosene produced by the processes outlined earlier.

Some approaches add hydrogen to the pyrolysis reactor. This is known as hydropyrolysis and

- lowers the production of heavy tars
- lifts the yield of liquids
- lowers the cost of further refining of the liquids.

165

Table 5.1. Classification of Pyrolysis Types and Approximate Product Yields[1]

Pyrolysis	Temperature (°C)	Vapour residence time	Heating rate (°C/s)	Feedstock size (mm)	Gas (wt. %)	Oil (wt. %)	Char (wt. %)
Slow	300–700	10–100 min	0.1–1.0	5–50	35	30	35
Intermediate	500–600	0.5–20 s	1.0–10	1–5	35	50	25
Fast	400–800	0.5–5 s	10–200	<3	30	50	20
Flash	800–1,000	<5 s	>1,000	<2	13	75	12

Approach to the Direct Conversion of Biomass to Fuels

When heated in the absence of oxygen, solids that contain organic matter (wood, biomass, coal, oil shale, etc.) will emit gases and condensable liquids. The solids remaining contain fixed carbon and can often be used as a fuel or, in the case of biomass, used as an organic carbon supplement for agricultural purposes. For several centuries, this process has been used to produce a wide range of gases, liquids and solids of benefit to humanity. Some technologies are

- Coal pyrolysis to produce coke, coal tar chemicals and town gas (gas works and steel coke ovens).
- Oil from oil shale.
- Wood pyrolysis to produce chemicals and charcoal.

These approaches can be used as the basis for the design of biomass pyrolysis units. Because these technologies have been used for centuries, there are many variants.[2] However, several generalisations can be made:

1. Unlike the indirect route (synthesis gas route), most technologies are developed around a particular solid, at a particular location, to produce a particular product. This means that it is often difficult to transfer a technology from one location to another without incurring additional development costs.

---

[1] Wikipedia, Wikimedia Foundation, *Pyrolysis,* downloaded June 24, 2024.
[2] The Ralph M. Parsons Company, "Coal Liquefaction Process Research — Process Survey," US DoE ORNL/Sub-7186/13, December, 1977.

2. The composition (quality) of the gases and liquid products is dependent on the feedstock composition, technology and conditions (temperature and pressure).
3. High-temperature pyrolysis generally results in liquid products that have a high level of unsaturated materials present, including aromatics. For the most part, liquid products require a significant amount of hydro-treatment to make a saleable product such as SAF.

Because of these issues, only general indications for the conversion of biomass can be given without a significant amount of further data being required. The results should only be considered for guidance only. The principal feedstock for these earlier process operations was low-rank coal and lignite. In recent years, these technologies have been adapted to use biomass as a feedstock. Some of the pertinent past technologies that could inform the development of biomass pyrolysis are

Shale Oil Production

The production of oil from shale has been known for several hundred years.[3] The general layout for an oil shale retort is shown in Figure 5.1.

Oil shale is crushed and passed via lock hoppers to the top of the retort where the oil shale is heated using fuel gas and air introduced lower down in the retort. Fuel gas combustion can be done inside the retort, directly heating the shale, or indirectly by combustion in an external jacket. Liquids and gases are expelled from the oil and separated. Part of the fuel gas stream is recycled for use in heating the retort. Solids (ash) fall to the bottom of the retort and are withdrawn from the base of the retort.

There are many variants on the scheme, mainly concerning the method of heating the retort. In one scheme (Union Process[4]), a solid lift arrangement caused the shale to ascend the retort counter-current to descending hot gases. In another process (Tosco[5]), ceramic balls are used

---

[3] "An Assessment of Oil Shale Technologies," US NTIS #PB80-210210115, June, 1980.

[4] Wikipedia, Wikimedia Foundation, *Union Process*, downloaded March 25, 2014.

[5] M.T. Atwood and B.L. Schulman, "The Toscoal Process Pyrolysis of Western Coal and Lignite for Char and Oil Production," in *Annual meeting, American Institute of Mechanical Engineers*, Denver, Colorado, February 15–19, 1970.

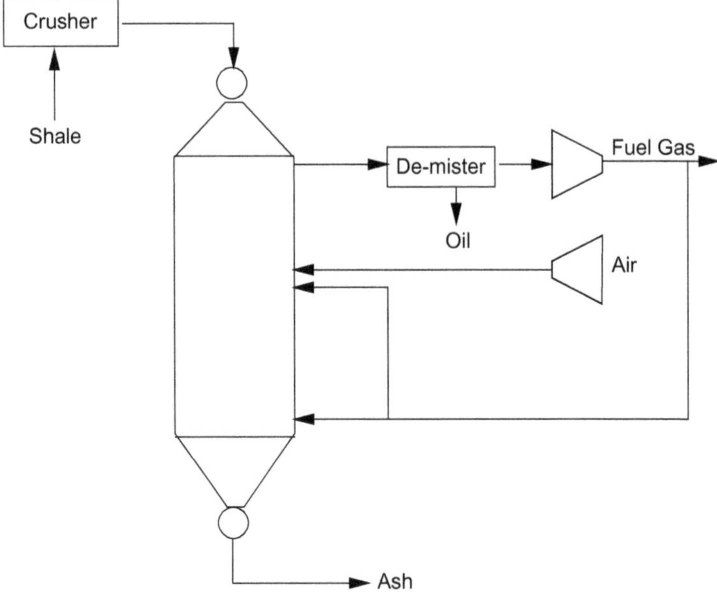

Figure 5.1. Oil Shale Retort

as a heat transfer agent. The balls are heated by combustion of the residual carbon in the shale. In other processes, hot ash is used.

Some specific technologies of interest are as follows:

Taciuk Process: Stuart Oil Shale

The Alberta Taciuk[6] processes oil shale fine particles in a rotating kiln. The unique feature of the Alberta Taciuk process is that drying and pyrolysis of the feed shale and the combustion, recycling and cooling of spent shale all occur in a single multi-chamber horizontal rotating vessel. The water pollution caused by this process is quite limited. Australian oil companies Southern Pacific Petroleum NL and later Queensland Energy Resources operated a 250-tonnes-per-hour industrial-scale pilot plant using the Alberta Taciuk Processor. The plant was shut down in 2004.

---

[6]Wikipedia, Wikimedia Foundation, *Blue Ensign Technologies,* downloaded June 13, 2024.

Rendall Process

In the newer Rendall process,[7] pyrolysis is accompanied by hydrogenation in a very similar flow sheet to coal hydrogenation, which is dealt with here. In the Rendall process, oil shale is heated in the presence of a hydrogen donor solvent (tetralin), which helps mobilise the organic component of the shale (kerogen). In the second step, liquids are obtained by supercritical extraction using toluene. Spent shale is separated, and the liquids are distilled to recycle heavy oils and the solvents (Figure 5.2).

Oil shale is ground (6 mm) and passed to a conditioning vessel where the shale is formed into slurry (30%–50% solids) with the solvent tetralin. The slurry is heated to 450°C and about 45 bar. Water, which produces superheated steam, is added. This causes the organic material present (the kerogen) to convert to liquids and more easily extracted materials. The products are separated into hydrocarbon liquids and gases and an oil slurry, which is passed to the solvent extractor. This removes oil from the solids, which is passed to the distillation section.

Distillation removes the oil product. Spent tetralin (as naphthalene) is passed to the solvent hydrogenation vessel and recycled to the slurry

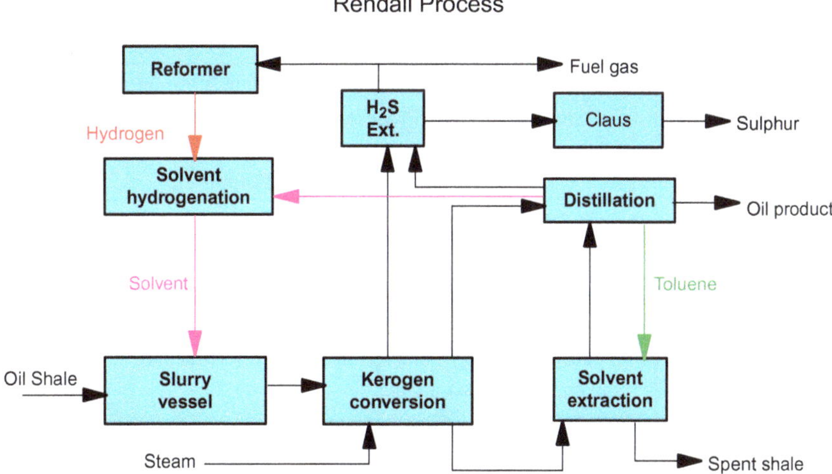

Figure 5.2. Rendall Process Block Flow

[7] Wikipedia, Wikiemedia Foundation, Investment Profile, *Blue Ensign Technologies Ltd.*, February 18, 2008.

vessel. Hydrogen sulphide (H_2S) is removed from the gas fraction, which is subsequently converted to sulphur. The hydrocarbon gases are sent to a steam reformer, which produces hydrogen for the solvent regenerator. Excess gas is used as fuel.

The solvent/pyrolysis process claims much higher yields of liquid than would normally be expected from an inspection of a Fischer assay of oil shale. The proponents claim a 150% to 300% increase in the liquid yield with 90% conversion of the carbon present into liquids and gases.

Other Processes

Table 5.2 compiles several process types that have been developed or are in the process of development.

Table 5.2.　Some Oil Shale Processing Technologies[8]

Heating method	Above ground (ex situ)	Below ground (in situ)
Internal combustion	Kiviter, Fushun, Union A, Paraho Direct, Superior Direct	Oxy MIS, LLNL RISE, Geokinetics Horizontal, Rio Blanco
Hot recycled solids (inert or burnt shale)	Alberta Taciuk, Galoter, Lurgi, TOSCO II, Chevron STB, LLNL HRS, Shell Sphere	
Conduction through a wall (various fuels)	Pumpherston, Hom Tov, Oil-Tech, EcoShale In-Capsule Process, Combustion Resources	Shell ICP (primary method), EGL Oil Shale Process, IEP Geothermic Fuel Cell Process
Externally generated hot gas	PetroSIX, Union B, Paraho Indirect, Superior Indirect, Syntec process (Smith process)	Chevron CRUSH, Petro Probe, MWE IGE
Reactive fluids	IGT Hytort (high-pressure H_2), Xtract Technology (supercritical solvent extraction), Donor solvent processes, Chattanooga fluid bed reactor	Shell ICP (some embodiments)
Volumetric heating		ITTRI, LLNL and Raytheon radiofrequency processes, Global Resource microwave

---

[8] Wikipedia, Wikimedia Foundation, *Shale Oil Extraction*, downloaded June 13, 2024.

Coal Pyrolysis

Coke Making

Technology for the production of coke has been known for many hundreds of years. There are many forms of the process; the two main ones are for the production of coke for iron manufacture and for the production of gas and chemicals.

Carbonisation refers to the heating of bituminous coal in ovens sealed from air to form coke. The process involves thermal decomposition of the coal with distillation of the products. Various technologies are used that perform the process at (i) low temperature (500°C–750°C), (ii) medium temperature (750°C–900°C) and (iii) high temperature (900°C–1,175°C). High-temperature operation generally favours the production of coke for ironmaking.

Coke Ovens and Retorts

The coke ovens are held in batteries of many ovens producing coke on a batch basis.

A typical coke oven is about 40 ft long and 14 ft high with an average width of 17 inch. The coke oven is tapered to facilitate coke removal. The ends of the oven are closed with removable doors. The oven is filled from charge holes at the top of the oven. Volatile products leave the oven from openings in the top, which transfer the volatile materials to collecting mains. The coke oven is heated with a coke oven gas burnt in the oven walls. Typically, 35% of the coke oven gas is used in this process.

When carbonisation is complete, a pusher machine pushes the hot incandescent coke out of the oven into a receiving truck.

For the production of town gas, the operation is similar. Carbonisation is usually performed at a lower temperature, and the ovens are smaller and generally referred to as retorts.

After production, the volatile matter is passed to a downstream processing train, which removes the products. This is similar for both coke ovens and gas retorts. The general layout is illustrated in Figure 5.3.

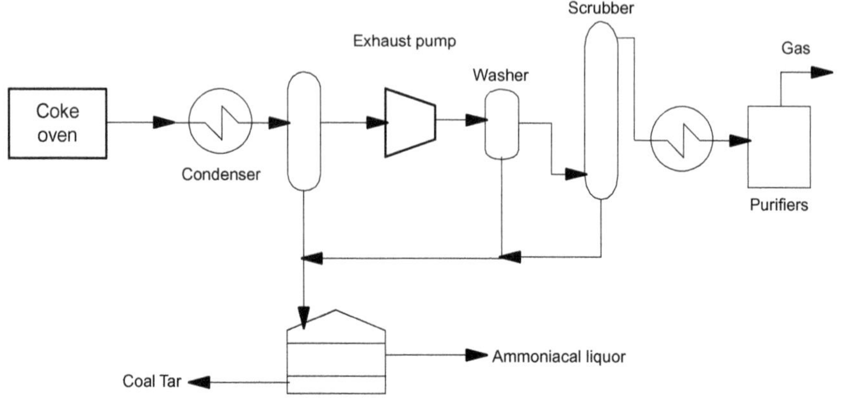

Figure 5.3. Coke Oven and Downstream Operations

Table 5.3. Typical Composition of Coal Gas

	Vol %
Carbon monoxide	6.8
Hydrogen	47.3
Methane	33.9
Carbon dioxide	2.2
Nitrogen	6.0
Ethylene etc.	3.8
Fuel value MJ/m^3	22.0

The coal tars are worked up to produce coal chemicals: naphtha, cresols and phenols. The ammoniacal liquor is distilled to produce ammonia (which is often converted into ammonium sulphate).

In carbonisation systems, coke forming in the absence of oxygen restricts the formation of carbon oxides to what can be formed from water and oxygen present in the coal. Some hydrogen is formed by the water–gas–shift reaction, but most (and the methane formed) is a consequence of the decomposition of the large coal hydrocarbons into the elements. The typical gas composition is shown in Table 5.3.

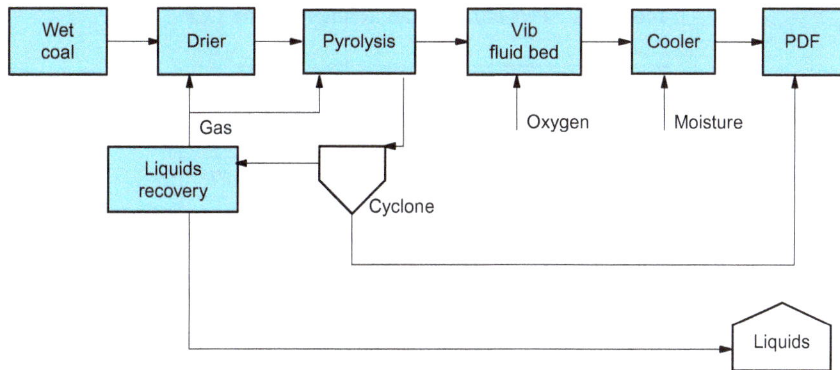

Figure 5.4. Unit Operations in LFC Process

Newer Approaches to Pyrolysis

Brief History

During the 1990s, ENCOAL and the US DoE demonstrated the SGI International "liquids from coal" (LFC) process.[9] The LFC process was aimed at improving the utility of low-rank coals of high moisture content to produce a high heat content solid (process-derived fuel [PDF]) and a liquid (coal-derived liquid [CDL]). The process block flow diagram is shown in Figure 5.4.

Wet coal (30% moisture) is dried and then heated in a pyrolysis unit to drive off the volatile matter. The solid product is passivated in a vibrating fluid-bed unit (VIB) in the presence of a small amount of oxygen. The solid is cooled, and moisture is added back to give the PDF. From 1 tonne of raw coal, 0.5 tonne of PDF and 0.5 bbl of liquid were produced. The properties of the coal and the PDF are as given in Table 5.4. The properties of the liquids (CDL) from the LFC process in comparison to low sulphur fuel oil (LSFO) are given in Table 5.5.

As typified by the LFC and coking operations discussed earlier, the principal problem facing the use of the pyrolysis route to the production of liquids for fuel use is the formation of coal tar, often referred to as

---

[9] "The ENCOAL Mild Coal Gasification Project," US Department of Energy DOE/NETL-2002/1171.

Table 5.4. Properties of the Solids from the LCF Process

Property	Feed coal	PDF
Heating value (BTU/lb)	8,400	11,200
SO_2 (lb/MMBTU)	1.1	0.7
Water (wt. %)	29	9
Ash (wt. %)	5	8
Volatiles (wt. %)	31	24
Fixed carbon (wt. %)	35	59
Sulphur (wt. %)	0.45	0.4
Ash fusion temperature (°F)	2,220	2,220

Table 5.5. Properties of Liquids from the LCF Process

Property	LSFO (for comparison)[a]	CDL
API gravity (°)	5	2.3
Sulphur (%)	0.8	0.6
Nitrogen (%)	0.3	0.7
Oxygen (%)	0.6	10.8
Viscosity (122°; cSt)	420	240
Pour point (°F)	50	80
Flash point (°F)	150	218
BTU/gal	150,000	140,000

[a]Low sulphur fuel oil

black tar. This material contaminates the liquid streams, making them extremely viscous (e.g. pour points of 90°F); this degrades the liquids to the point that they can only be used as a substitute for heavy fuel oil.[10]

The formation of black tar is explained as a consequence of the polymerisation of the highly unsaturated pyrolysis products catalysed by particle fines and oxygen. One approach to this problem is various methods

---

[10]Discussion in A.P. Fraas, R.L. Ferguson and H.L. Falkenberry, *US Patent* 7,008,459, March 5, 2006.

and converter designs using steam as a heat transfer reactant and to help gasify the produced tars.[11]

Another approach to solving the problem is to perform the pyrolysis in the presence of hydrogen. This would be expected to hydrogenate the pyrolysis products and therefore prevent polymerisation. Most of the newer approaches involve some variation of hydropyrolysis. For example, Ikura and Last[12] describe a flash hydropyrolysis process using bituminous coals. At about 700°C, 63% conversion of the coal is claimed with a selectivity of

Organic liquids	5.1%
Process water	10.5%
Gaseous hydrocarbons	31.3%
CO and CO_2	5.5%
Char	44%

The problem with the pyrolysis processes discussed earlier is that substantial energy input is required in order to induce the pyrolysis. One way to effect heat input is to partially combust part of the coal in the reactor. Such a system is then a mix of pyrolysis and gasification. The Nippon Steel process is of this type and will be described.

Nippon Steel Pyrolysis Process[13]

The process is shown diagrammatically in Figure 5.5.

Coal and oxygen enter the partial oxidation unit (POX). Slag falls through the slag trap into the base of the reactor. The hot gases ascend into

---

[11] For example, H. Kubiak, H.J. Schroter, G. Gappa, H. Kalwitzki, K. Knopp, *US Patent* 5,064,444, November 12, 1991, to Bergwerksverband Gmgh details allothermic generators using super-heated steam as a heat transfer and gasification agent. This invention claims that with a coal feed rate of 60 t/h, a steam input rate of 2.67 t/t coal at 900°C, 95% coal conversion can be achieved. This requires a heat input of 340 MW (this will require a similar coal rate to a steam boiler). This is essentially a full steam gasification system with the principal products being synthesis gas. It is an alternative to a gasifier.

[12] M. Ikura and A.J. Last, *US Patent* 4,935,036, June 18, 1990, to Energy Mines and Resources — Canada.

[13] H. Yabe *et al*, "Development of Coal Partial Hydropyrolysis Process," *Nippon Steel Technical Report*, July, 2005.

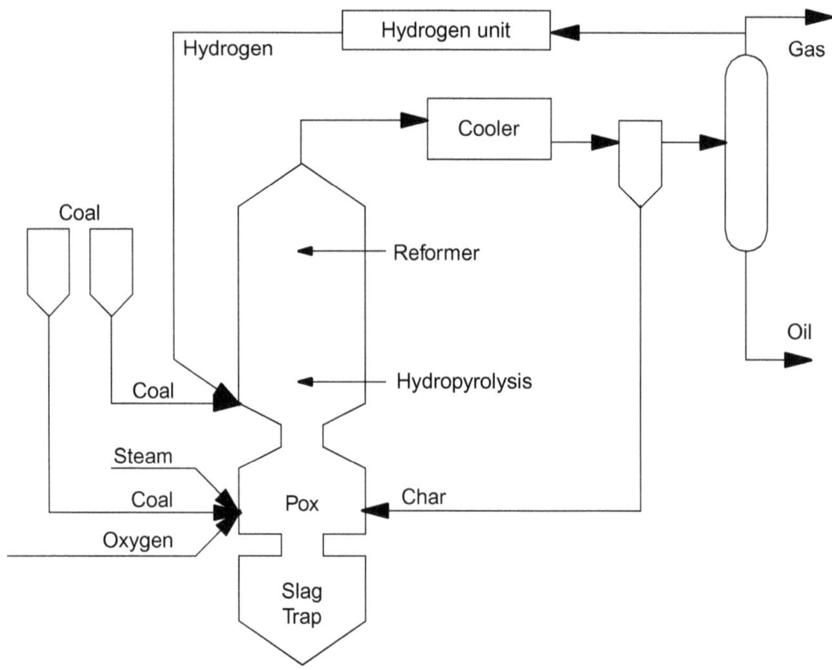

Figure 5.5.   Nippon Hydropyrolysis Process

the hydropyrolysis section of the unit. Here, coal and hydrogen are heated to the pyrolysis temperature where the breakdown of the coal occurs. The hot gases ascend into a reforming section where more coal is added with extra hydrogen. Here, hydropyrolysis and hydrogenation of the heavy pyrolysis tars to lighter oils occur. The products are cooled, char is removed and recycled to the POX unit, and the products are cooled and separated into a gas fraction and a light oil fraction. Part of the product gas is sent to a hydrogen unit, which produces hydrogen for the hydropyrolysis section. A small amount of nitrogen enters the system as a carrier gas for the coal streams.

Chemistry

The chemistry in each section of the unit is as follows:

Partial oxidation: In this section, coal is partially burnt:

$$CH_mO_n \text{ (coal)} + O_2 + H_2O = CO + H_2 + CO_2 + H_2O + C(\text{char}).$$

Hydropyrolysis: In this section, several reactions occur:

Pyrolysis: Coal -> Gas (CO, CO_2, H_2, C_mH_n, H_2O) + Tar + BTX + Char.
BTX is benzene, toluene and xylene. This section also includes the cracking of large molecules by free radical processes to form smaller moieties. The free radicals are quenched by hydrogen.

Radical termination: Free radical + H_2 = Gas + Tar
Char gasification: C (char) + H_2O = H_2 + CO + CO_2

Reformer: In this section, hydrogenation of heavy tar occurs:

$$Tar + H_2 = C_mH_n(gas) + BTX + oil.$$

The process is designed to operate at a pressure of 2 to 3 MPa. As the gases rise in the reactor, the temperature gradually falls. The temperature in the POX is 1,500°C to 1,600°C falling to 700°C to 900°C in the reformer.

Table 5.6 gives details of the estimated inputs and outputs for a full-scale plant taken from Nippon Steel data.

The hydrogen and the char are recycle streams within the unit. The total output has been made to match inputs by assuming that 1% of the gas stream has a molecular weight of pentane.

What is striking about the results is that most of the product mass is in the gas stream. Oil is only a minor product. The gas analysis for the Nippon process on a dry basis is

Methane	8%
Hydrogen	29%
Carbon monoxide	49%
Carbon dioxide	6%
Nitrogen	7%
Others	1%
LHV	13 MJ/cm

The relative amount of gases and liquids produced seems to be dependent upon the temperature in the reformer section, with lower temperatures favouring higher liquid yields. Table 5.7 illustrates the point.

Table 5.6. Nippon Steel Pyrolysis Data

Partial oxidation		
Coal to POX unit	t/h	15
Oxygen	Ncm/h	14,000
	t/h	18.96
Steam	t/h	3
Char	t/h	10
Total	t/h	46.96
Hydropyrolysis		
Coal to hydropyrolysis	t/h	21
Hydrogen	Ncm/h	1,300
	t/h	0.11
Total inputs		57.96
Outputs		
Gas	Ncm/h	61,000
	t/h	53.94
Oil	t/h	3
Slag	t/h	1
Total outputs	t/h	57.94

Reading from graphs, the yields of the principal fractions compared to coke oven operations are approximately as given in Table 5.8.

The small amount of process water produced enters the product gas stream. For most downstream use, this will be removed from the gas stream by condensation and drying. There should be no excessive contamination of this water. Clean water is required for the production of boiler feed water for raising steam.

As illustrated earlier, the major product of the process is gas. This requires disposal in a downstream process in much the same manner as the full gasification approach.

The Nippon Steel process is said to be under demonstration on a 25 t/d plant. This phase was aimed to have been completed in 2008.

Table 5.7. Approximate Liquid Yields at Different Reformer Temperatures

Reformer temperature (°C)	700–800	900–950
Carbon conversion to gas	42%	70%
Carbon conversion to liquids	20%	15%
Carbon conversion to BTX	4%	5%
Carbon to char	37%	18%

Table 5.8. Approximate Liquid Yield According to Fraction

Boiling range	Nippon technology	Coke oven
<220°C; gasoline and kerosene	70%	15%
220°C–360°C; diesel	25%	40%
>360°C; residua	5%	45%

Pyrolysis of Biomass

The properties of the products and compositions using biomass as opposed to coal as feedstock are expected to be similar in character to the products produced from coal or lignite.

To some extent, the properties of bio-oil are dependent on the feedstock, temperature of pyrolysis and other process conditions. However, there are some general characteristics of the oil that are relevant to its use.

In some form, the bio-oils are generally free-flowing with a dark reddish-brown colour. They have a sharp, smoky odour reflecting the presence of acids and aldehydes. The density is very high compared to fossil fuels, often with a density of more than unity (e.g. 1.2 kg/L). This is indicative of the presence of suspended solids and large aromatic molecules (asphaltenes). Bio-oils have high viscosities (40–100 cP). Relative to petroleum fuels, the heating value is relatively low (e.g. 27 MJ/kg) indicating a high oxygen content.

On standing, the viscosity rises, various phases can be separated and gum can be deposited. This is indicative of slow polymerisation initiated by the presence of free radicals and unsaturated molecules, which are readily polymerised. This raises issues with the transport of these materials to a facility able to refine these oils.

The bio-oils are immiscible with non-polar solvents but are miscible with polar solvents.

Bio-oils, like coal products, contain oxygenated molecules such as phenols and cresols (both with densities greater than unity) as well as hydrocarbons such as benzene, toluene and xylenes (BTX) and aliphatic hydrocarbons. Water phases contain acids and aldehydes.

Role in SAF Production

The main interest arises from the presence of high-density aromatic rings in the pyrolysis liquids and tars. This has to be worked up by selective hydrogenation to produce the pure mononuclear aromatics, BTX. For example, the reduction of cresol to toluene:

$$CH_3.C_6H_4.OH + H_2 = CH_3.C_6H_5 + H_2O.$$

These mononuclear aromatics can be used to produce an aromatic containing a longer paraffinic chain by the addition of a suitable olefin. For example, using pentene to produce amyl toluene:

$$CH_3.C_6H_5 + C_5H_{10} = CH_3.C_6H_4.C_5H_{11}.$$

Such processes are well known in the petrochemical industry for the production of specialist chemicals and detergents. The amyl toluene has a high density (0.9 kg/L) and other properties (b.p. 228°C and flash point 86°C), which should be acceptable for jet fuel blending with low-density paraffinic SAF. The products produced by such processes are commonly referred to as sustainable aviation kerosene (SAK).

Economics of SAK Production by Biomass Pyrolysis

There is limited data concerning the production of SAK by pyrolysis. As noted earlier, the pyrolysis of biomass produces three products — gas, oil and char. Char is seen as a valuable agricultural supplement (fertiliser), and some data is available from studies on the production of biochar by pyrolysis.[14] The case study developed here uses an *a priori* method for estimating the cost of SAK production.

The case takes place in three steps: (i) the production of oil and char, the gas fraction is used as fuel for the facility, (ii) the upgrading by hydrogenation of the bio-oil and (iii) the production of SAK.

The Production of Bio-Oil and Biochar

Guidance for this cost estimation is taken from the production costs for oil shale. One million tonnes of biomass is subject to pyrolysis in a similar manner to that for shale oil retorting (Figure 5.1). Figure 5.6 illustrates the process modelled, and the inputs and outputs are detailed in Table 5.9.

The biomass is stored and dried before entering the retort. The product gases, liquids and solids are separated with the gas fraction used to fuel the pyrolysis. Solid material (char) is extracted for sale. The crude oil is stabilised, and the bio-oil is passed on to downstream operations for upgrading and refining.

Relative to shale oil, the bio-oil has a high density, and it is assumed that water produced in the pyrolysis is dissolved in the oil phase.

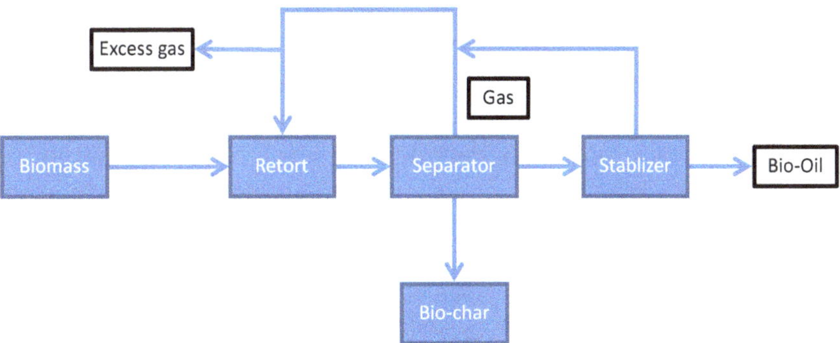

Figure 5.6. Pyrolysis of Biomass to Produce Bio-Oil and Bio-char

---

[14]T.R. Brown, M.M. Wright and I.C. Brown, "Estimating Profitability of Two Biochar Production Scenarios: Slow Pyrolysis vs. Fast Pyrolysis," *Biofuel. Bioprod. Biorefin.* **5**(1), 54–68 (2011).

Table 5.9. Pyrolysis of Biomass

Biomass	Mt/y	1
	kg/h	122,549
Oil	Mt/y	0.650
	kg/h	79,657
At oil density of 1.2 kg/L	bbl/d	10,027
Gas	kg/h	24,510
Char	kg/h	18,382
Total out	kg/h	122,549

The composition of the products is taken as 20% gas, 65% oil and 15% char. This is similar to the composition for a fast/flash pyrolysis method.

The order of magnitude capital costs is given in Table 5.10.

Since maximum oil yield would be delivered by fast or flash pyrolysis, the capital cost of the retorts may be higher than given in Table 5.10. The estimated non-feedstock operating costs are given in Table 5.11.

To minimise transport and storage problems with the unstable bio-oil, the pyrolysis facility is assumed to be juxtaposed to the downstream upgrading operations. For this scale of the operation, it is assumed that biomass feedstock is available from several suppliers (farmers) so that an economy of scale can be realised. The biomass is assumed to be agricultural waste (corn stover, rice hulls, almond shells etc.) with no value and that the base cost used in the estimate is the cost of transport to the facility at $10/t. The estimate of the crude bio-oil production cost is given in Table 5.12.

The crude oil produced is approximately 10,000 bbl/d; the crude oil has a very high density of 1.2 kg/L. There is ample gas produced to power the pyrolysis section of the plant. This is assumed to be burnt in the pyrolysis process and to raise steam for power generation. Excess power is exported, generating a credit of $6.14 M/y.

The pyrolysis also produces 150 kt/y of biochar. The value of this has been estimated at $50/t, which is below current (2024) prices (of about $100/t), and allows for the cost of upgrading the product to an acceptable standard for agricultural use. No account has been taken for the generation

Table 5.10. Estimated Capital Costs for Biomass Pyrolysis to Crude Bio-Oil

CAPEX		
Biomass preparation	$M	38.18
Retorts	$M	69.69
Gas turbine and steam	$M	8.04
Cooling water	$M	8.60
Subtotal	$M	198.72
Civil engineering	$M	20.65
Piping	$M	15.94
Instrument and control	$M	5.84
Electrical	$M	17.96
ISBL	$M	259.09
Owners	2% of ISBL	5.18
Engineering	10% of ISBL	25.91
Contingency	25% of ISBL	64.77
Royalty	2% of ISBL	5.18
Catalyst and chemicals	2% of ISBL	3.89
Start-up	1% of ISBL	2.59
Spares	2% of ISBL	5.18
Land	$M	0.20
Subtotal	$M	112.91
Total CAPEX	$M	372.00

Table 5.11. Estimate Operating Cost of Biomass Pyrolysis to Crude Bio-Oil

Cooling water (5 c/L)		0.12
Chemicals	1.50% of ISBL	3.89
Catalysts		0.50
Maintenance	3.50% of ISBL	9.07
Insurance	1.50% of ISBL	3.89
Labour	2.00% of ISBL	5.18
Total OPEX	$M/y	22.64

Table 5.12.   Statistics to Produce Crude Bio-Oil

Biomass input	kg/h	122,549
Biomass cost	$/t	10
	$M/y	10
Oil output	kg/h	79,657
	bbl/d	10,027
Export power	MW	15.06
Power value	$/MWh	50
	$M/y	6.14
CAPEX	$M/y	372.00
ROC (14.85% Capex)	$M/y	55.28
Return on working capital	$M/y	0.32
Capital costs	$M/y	55.60
OPEX	$M/y	22.64
Costs	$M/y	88.24
Char credits	$M/y	7.50
Power credits	$M/y	6.14
Credits	$M/y	13.64
Net costs	$M/y	74.60
Unit production cost	$/bbl	21.88

of carbon credits, which may arise from the use of biochar. The sale of the char generates a credit of $13.6 M/y.

These statistics generate a crude oil production cost of about $22/bbl. This low value is reflective of the low feedstock cost for this route. However, this crude bio-oil requires substantial refining to produce a product suitable for transport fuel use and SAF. The cost of this upgrading adds significantly to the cost of production.

Refining of Crude Bio-Oil to SAF (SAK)

The possible process steps we might expect for upgrading the crude bio-oil are figuratively displayed in Figure 5.7.

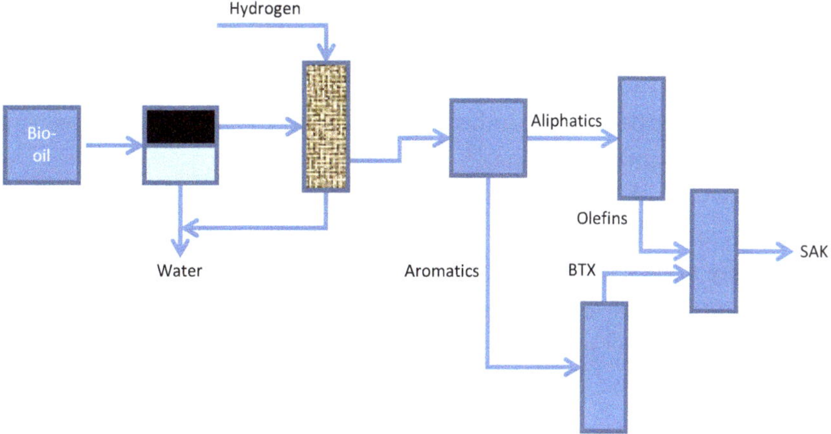

Figure 5.7. Possible Unit Operations in Upgrading and Refining of Bio-Oil to SAK

Stabilised bio-oil from the pyrolysis section is treated to separate out the water. This process is expected to be similar to the treatment of petroleum crude oil containing water as an emulsion. At its simplest, this may be by the application of heat to break the emulsion, but more elaborate technologies are available.[15]

The oil is passed to a hydrotreater unit with the duty to remove the oxygen from the bio-oil and produce a hydrocarbon product. The assumption is that a refinery-style hydrotreater will be able to accomplish this task.[16] However, the bio-oil is so heavily laden with oxygenate materials that a standard hydrotreater will likely require modification. In particular, the unit will have to handle much higher heat evolution from the hydrogenation processes, and special catalysts may be required. Furthermore, the process would require a significant amount of hydrogen to remove all of the oxygenates present. For the purpose of this estimate, it is assumed that the oxygen present is as phenol compounds so that each oxygen atom requires one hydrogen molecule (H_2) for removal.

---

[15] K.R. Albinson, C.H. Hsai, T.R. Melli and G.L. Woolery, *Microwave Separation Technology (MST)*, AIChE, Spring Meeting, 2000 and *Hydrocarbon Processing*, November 2000, p. 138.

[16] R.A. Meyers, *Handbook of Petroleum Refining Processes*, 4th ed. McGraw Hill, 2016.

Following hydrogenation and oxygen removal, it is assumed that the oil composition is mainly aromatics (BTX) and aliphatics (paraffins and olefins). These can be separated using any one of a variety of processes used in the petrochemical industry. Extraction of aromatics using sulpholane absorbent is common.[17]

Following this, a group of operations is used, which are based on the petrochemical production of linear alkyl benzene, which are major commodity chemicals used in the production of detergents.[18] In this process, the aliphatics are dehydrogenated to olefins that are then reacted with benzene. In the petrochemical process, toluene and xylenes are demethylated to benzene. This step may not be required for the production of SAK.

Statistics for the process are given in Table 5.13.

Table 5.13.   Estimate for the Production Cost of SAK from Bio-Oil Feedstock

SAK	t/y	487,500
	bbl/y	3,652,572
ISBL Capex	$M	473.94
CAPEX	$M	789.89
Return on capital		14.85%
	$M/y	117.30
Working capital 30 days	$M	22.56
Return on WC	$M/y	2.256
Bio-oil	bbl/d	7,520
	$/bbl	21.88
	$M/y	55.95
OPEX	$M/y	
Labour	2% of Capex	15.80
Maintenance	3.50% of Capex	27.65
Catalysts and chemicals	1.50% of Capex	11.85
Others	2.00% of Capex	15.80

---

[17] *Hydrocarbon Processing* "2021 Petrochemical Processes Handbook.
[18] Idem.

Table 5.13. *(Continued)*

Subtotal		71.09
Hydrogen costs at $3,000/t		114.3
Total cost	$M/y	360.85
By-products	$M/y	0
Net costs	$M/y	360.85
	$/bbl	98.79

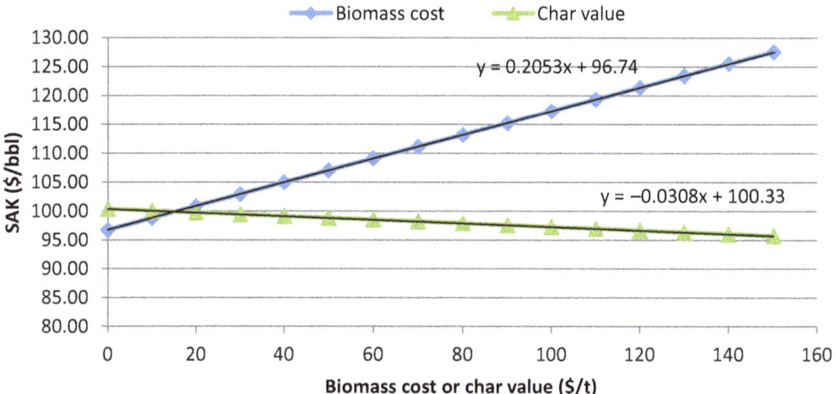

Figure 5.8. Sensitivity of Final SAK Production Cost to Biomass Cost and Char Value

The ISBL capital cost ($M 474) is based on the reported capital costs of the underlying processes discussed earlier. This estimate of under $100/bbl for the product of sustainable kerosene is much lower than the earlier estimates for SAF. However, as noted earlier, this is a multi-step process, each of which will be subject to a fair degree of uncertainty. The following figures illustrate the sensitivity of the SAK production cost to various key variables.

Figure 5.8 illustrates the sensitivity of the estimated SAK production cost to the pyrolysis part of the process to the biomass cost and the biochar sale value.

The cost of the biomass in the case study is a low value of $10/t, which represents the transport cost from what is essentially a free (waste) biomass. It is highly likely that this will be available in only a few instances. If biomass is to be sourced from planted crops (which would include forest plantations and the like), then the cost will be considerably higher. At \$120/t, the estimated cost of the SAK rises to \$120/bbl, which is broadly in line with petroleum kerosene with oil at \$100/bbl.

The value of the biochar credit in the case study is placed at \$50/t. This is lower than current-traded prices for agricultural biochar and allows for some additional costs for producing a purified product — complete removal of phenol and cresol contaminants, for instance. If the biochar value is \$120/t, the estimated production cost of SAK falls marginally by about \$2/bbl.

The sensitivity to the capital cost of the pyrolysis section to produce crude bio-oil is shown in Figure 5.9.

The capital cost of the pyrolysis section in the case study is estimated at \$372 M. At the time of writing, there are no significant facilities that are performing the task of pyrolysing biomass to bio-oil. This capital cost estimate could be subject to a fair degree of error. Should a more realistic cost be double this value, then the production cost of the SAK would rise to about \$110/bbl.

As noted earlier, the cost of production of SAK is dominated by the second stage of the overall process, namely, the upgrading of the crude bio-oil to SAK. Again, there is no data readily available for this section,

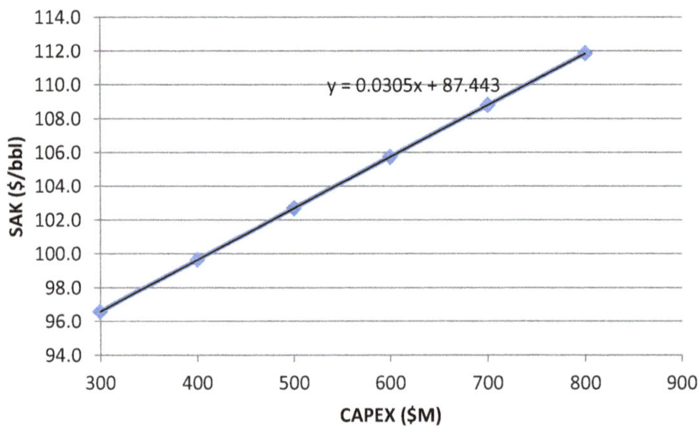

Figure 5.9.   Sensitivity of Final SAK Production Cost to Capital Cost of the Pyrolysis Section

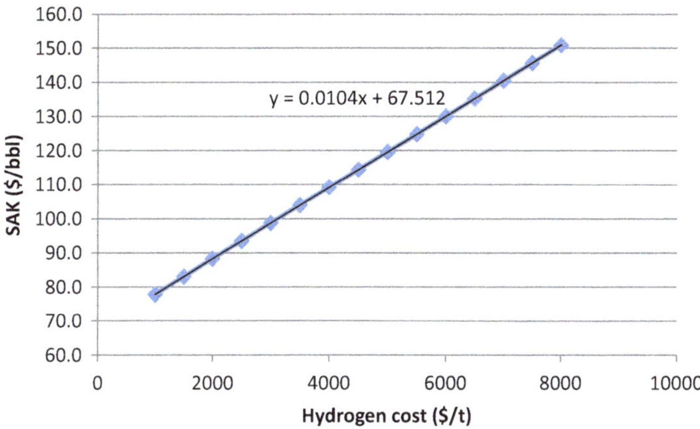

Figure 5.10. Sensitivity of Final SAK Production Cost to Cost of Hydrogen in the Second Stage

and so, the estimates in the case study could be subject to a fair degree of error. The process is highly dependent on the cost of hydrogen. The sensitivity to the cost of hydrogen on the production cost of SAK is shown in Figure 5.10.

The case study used a hydrogen cost of $3,000/t. This is similar to hydrogen costs in conventional oil refineries, where hydrogen is produced by steam methane reforming. If green hydrogen is to be used, then typically this is produced by the electrolysis of water. For low-cost hydrogen ($3,000/t), this requires very low electricity prices typical of power produced by hydropower operations. Higher electricity costs result in considerably higher hydrogen costs.[19] If the hydrogen cost is $6,000/t, then the production cost of the SAK rises to over $130/bbl.

---

[19]D. Seddon, *The Hydrogen Economy*, World Scientific, 2022.

6

COMMENTARY ON BLENDING AND COSTS AND OBSERVATIONS

Blending to SAF

In the previous chapters, I have discussed the technology and economics for making hydrocarbon components suitable for the manufacture of sustainable aviation fuel (SAF). None of the components are suitable in their own right for SAF. The main problem is overcoming too low a density for most of the products, whilst higher-density materials such as sustainable aviation kerosene (SAK) will deliver too high a level of aromatics for use in their own right. However, blends could be made to deliver SAF.

At the time of writing, regulatory authorities limit the use of these alternative sustainable components in jet fuel blendstock with conventional petroleum kerosene to produce an acceptable aviation fuel. The petroleum kerosene is typically the major component of the blend materials.

The American Society for Testing Materials (ASTM) has certified several of the sustainable components under the ASTM Method D7566[1] as suitable blendstock for aviation kerosene as detailed in Table 6.1.

---

[1] BP website, "How All Sustainable Aviation Fuel (SAF) Feedstocks and Production Technologies Can Play a Role in Decarbonizing Aviation," downloaded June 20, 2024.

These routes, that is, two technology pathways, are currently certified by the ASTM to produce SAF through co-processing in conventional refinery operations. These are as follows:

- Co-hydroprocessing from Fischer–Tropsch products to a maximum of 5% in the final product.
- Co-hydroprocessing from hydroprocessed esters and fatty acids to a maximum of 5% in the final product.

The development of these lists is ongoing, but clearly at the time of writing, these rules limit the production of a truly SAF.

The information from the BP website goes on to discuss the various options and position of the production of SAF. This view summarises the position developed earlier in this book, and I give BP's views here in verbatim:

Each of these feedstocks uses a particular production technology, with each specific technology pathway needing approval from the fuel standard body ASTM before being commercially deployed.

There are two ways of producing SAF, with standalone units or through co-processing. Standalone units use sustainable feedstocks to produce the synthetic kerosene (SK), that is then blended with conventional jet fuel to produce SAF. Whilst when producing SAF through co-processing, up to 5% sustainable feedstocks being are processed alongside fossil feedstocks through hydro-processing in the refinery.

With standalone units the feedstock is converted in a biorefinery into SK and then certified to the relevant annex in ASTM D7566 standard. The approved technology pathways and associated feedstocks are shown in Table 6.1. This SK is then blended up to 50% with conventional jet fuel and certified to ASTM D1655 or Defence Standard 91-091 and is supplied as a conventional Jet A/Jet A-1 fuel.

While HEFA synthetic paraffinic kerosene (SPK) is currently the only commercial pathway being used at scale to produce SAF, current feedstocks are limited. There is need for rapid commercial large-scale sustainable feedstock mobilisation. Alternative high energy crops that are being trialled or have already been approved as HEFA feedstocks, include algae, camelina, pennycress, tallow tree and carinata. We

Table 6.1. Technology Pathways Currently Certified by ASTM

Technology	Common name	D7566 Annex	Feedstock	Max % in blend	See Chapter(s)
Fischer-Tropsch hydroprocessed synthesized paraffinic kerosene	FT-SPK	A1	Waste CO_2 and renewable power. Municipal solid waste	50	4
Synthesized paraffinic kerosene from hydroprocessed esters and fatty acids	HEFA SPK	A2	Vegetable oils and waste oils (used cooking oils)	50	3, 5
Synthesized iso-paraffins from hydroprocessed fermented sugars	SIP	A3	Fermentable sugars	10	2
Synthesized kerosene with aromatics derived by alkylation of light aromatics from non-petroleum sources	FTSPK/A	A4	Waste CO_2 and renewable power. Municipal solid waste. Agricultural waste/waste wood	50	4, 5
Alcohol to jet synthetic paraffinic kerosene	ATJ-SPK	A5	Ethanol and isobutanol	50	2
Synthesized kerosene from hydrothermal conversion of fatty acid esters and fatty acids	CHU	A6	Vegetable oils and waste oils (used cooking oils)	50	3
Synthesized paraffinic kerosene from hydrocarbons esters and fatty acids	HC-HEFA SPK	A7	*Botryococcus brounii* species of algae	10	3, 5

promote the use of cover crops (crops grown for the protection and enrichment of the soil), such as carinata, as a feedstock when they do not require additional land demand and contribute to sustainable farming practices — supporting soil carbon accumulation, soil quality and biodiversity....

While an increasing number of flights have been fuelled by SAF produced from the HEFA pathway, limited feedstocks mean we expect to see SAF produced from alcohol to jet (AtJ), Municipal Solid Waste (MSW) and second generation (2G) biomass increasing significantly beyond 2030.

First generation alcohol to jet (AtJ)

Alcohol to jet (AtJ) is another technology that has an approved pathway. It is a method whereby sugary, starchy biomass such as sugarcane and corn grain are converted via fermentation into ethanol or other alcohols which can then be shipped or piped before being converted to fuel. These feedstocks are easy to grow and transport by train, however sugarcane must be processed into ethanol within 48 hours of being cut. To achieve low logistical costs, reduce carbon emissions from transport and make better use of infrastructure, ethanol plants benefit from being placed close to feedstock production mills as well as to refineries.

In some regions, particularly in the Americas, feedstock such as corn and sugarcane are currently commercially used for fuel production. Demand from sectors such as ground fuel and petrochemicals means however that there is limited feedstock available to aviation. As a result, there are no commercial SK plants using the AtJ production pathway.

Timing is therefore the important factor to consider with the AtJ pathway. As ground fuels move more towards electrification this will free up feedstock supply for the aviation sector which in turn will lead to commercial SAF production from this pathway.

Another consideration with the AtJ pathway is that the reduction in carbon intensity is not as strong when compared to the alternative technologies. Implementation of solutions such as carbon capture and storage technology will be key to lowering greenhouse gas emissions (GHG) using this technology. Other options to evaluate include the use of biogas in place of natural gas in mills and converting farm machinery to run on biofuels rather than fossil fuels.

Fischer-Tropsch (FT) for municipal solid waste (MSW)

As with AtJ there are approved pathways for this production. There is also feedstock availability, therefore producers are focusing on technology developments that can reduce the relatively high capital costs.

For SAF produced from MSW using Fischer-Tropsch (FT) technology the main environmental gain is derived from the fact that the waste would otherwise be left to decompose in landfill sites. According to the World Bank, the world generates more than 2 billion tonnes of MSW annually; a figure that's expected to grow to 3.4 billion tonnes by 2050. However, while access to MSW as a feedstock is widely available across the globe and it is typically a lower cost feedstock than other raw materials, in some regions aviation is in competition with other sectors, including the energy industry, for access to MSW.

In the EU, bp is — among other feedstock and technology pathways — advocating for recycled carbon fuels made from the non-organic portion of MSW to be recognised as SAF under the EU's planned SAF blending mandate. We are not however advocating for the use of recycled plastics as a standalone feedstock source for SAF.

It's a capital-intensive process to get the infrastructure in place, as production includes producing an FT wax which is then refined into SK before being blended into SAF. The good news here is that there is ongoing work to research and develop technologies that will lead to more efficient production. For example, bp and Johnson Matthey co-developed and co-own a simple-to-operate and cost-efficient FT technology that can operate both at large and small scale to economically convert synthesis gas, generated from sources including MSW, into long-chain hydrocarbons suitable for the production of SAF.

Technologies to convert 2nd generation biomass

For these feedstocks there is no pathway that is commercially deployed. However, work is progressing with ASTM for pyrolysis of biomass through both standalone production and co-processing in refineries. Once a pathway is approved, demo plants would then need to be established to prove the technology at scale before it is commercially deployed.

While second generation biomass such as agricultural and forestry residue is in vast supply as a feedstock, once aggregated it must be transported by road or rail — as a solid it can't be moved via pipeline.

Ultimately this means we'll end up with small production plants near where forestry residues are processed. While a handful of these plants already exist, none are being used currently to produce SAF.

Fischer-Tropsch (FT) for power-to-liquid

Possibly one of the most promising pathways for SAF in the longer term is power-to-liquid (PtL) technology (producing what is called eSAF), which is still very much in its infancy. Renewable electricity (from sources such as solar, hydro or wind) is used in an electrolysis process to extract hydrogen from water. This green hydrogen is first used to convert carbon dioxide (from the air, biogenic or industrial sources) to carbon monoxide. Then using FT synthesis technology, this carbon monoxide along with more green hydrogen is converted into a wax that can be upgraded to SK.

The challenge currently with eSAF technology is cost. To be commercially viable and competitive with conventional jet fuel this fuel (which is expected, in the short-term, to be three to eight times the cost of conventional jet fuel) needs to be produced at low cost. The availability and cost of the renewable energy and carbon dioxide, as well as the expansion and improvement of green hydrogen plants must be addressed to meet market demand.

Carbon dioxide could be secured from existing industrial sources such as plaster, concrete and food manufacturing companies as well as direct air carbon capture (a technology which removes carbon dioxide from the atmosphere). While sectors such as the steel and cement industries continue to explore technical solutions to reduce the carbon they emit, eSAF producers can benefit from using this cost-effective carbon dioxide source.

In terms of increasing green hydrogen production, as it is required by many industries to decarbonize, aviation will benefit from the focus being placed on increasing production. In addition, German mandates have specific requirements for eSAF from 2026 and European mandates currently being finalised are expected to follow suit in 2030.

Longer-term, eSAF will also benefit from on-going work in developing new pathways in addition to those already approved. In particular, methanol to jet could provide a competitive alternative production method for these feedstocks.

In summary, aviation is one of the hardest-to-abate sectors when it comes to reducing fuel lifecycle carbon emissions, with SAF currently the only way to decarbonize the industry at pace and at scale. Utilising a wide range of feedstocks is key to the production of SAF, as is the ongoing evolution of production pathway options. bp will continue to work with stakeholders across the energy supply chain as well as governments, NGOs, authorities and other businesses to help meet future SAF demand.

Discussion on the Estimated Cost of SAF

In the previous chapters, I have developed the underlying cost structure for the production of SAF from various feedstocks. SAF is a hydrocarbon, and so, the ultimate challenge for a feedstock that is competitive with conventional aviation fuel is to identify sources of low-cost carbon. Table 6.2 gives an indication of the cost of carbon from pertinent feedstocks.

The cost of carbon from conventional fuels is approximately $1,000/t. Sucrose and ethanol have a similar cost, but the conversion of these

Table 6.2. Estimates for Cost of Carbon for Feedstock

	Basis	Carbon (wt.%)	Feedstock cost ($/t)	Carbon ($/t)
Crude oil ($100/bbl)	7.4 bbl/t	83.0	740	892
Gasoline	8.6 bbl/t	87	906	1,041
Kerosene	$C_{12}H_{26}$	84.7	881	1,040
Diesel	7.43 bbl/t	86	833	969
Ethanol	C_2H_5OH	52.2	500	958
Sucrose	$C_{12}H_{22}O_{11}$	42.1	450	1,069
Palm oil		62	800	1,290
Rapeseed		62.1	1,400	2,254
Wood	Pine	47.5	150	316
Biomass	Corn stover	35	10	29
Municipal waste		45	30	67

materials to jet fuel is somewhat complex when compared to large-scale refining operations that have been honed to high efficiency over the years.

Palm oil is a mass-produced commodity that is widely used for the production of biofuels. The cost of carbon at a typical traded price is about 20% higher than that for petroleum fuels. Rapeseed and similar vegetable oils (canola) have high costs because they have other high-value uses, such as food and food processing. This translates into a high cost for the carbon content. If the vegetable oil route to SAF is chosen, then feedstock should be chosen from materials that are produced in excess of the market demand for other uses, for instance, tallow.

Biomass, including wood and municipal waste, has a much lower price for carbon, but the SAF production cost is very high compared to alternative approaches. This is because, with the current technology, the overall process efficiencies are low.

In general, the approaches from renewable fuels involve several steps. Several of these unit operations produce an intermediate that could be sold into an alternative market. Using sucrose as a feedstock gives the following intermediates with alternative use:

- sucrose itself is used in food production
- ethanol intermediate for use as a fuel additive and octane booster
- ethylene intermediate for use in chemicals and plastics manufacture
- alpha-olefins intermediate for use in detergent manufacture.

Marking the opportunity value at the intermediate stage inevitably lifts the final SAF production cost. To avoid this, full integration of the facility from feedstock to SAF would hinder the sale of intermediates and lower the overall production cost of SAF.

The application of government grants and tax incentives currently available for land transport fuels may not be available for SAF. It may be argued that they should apply for purely domestic operations, but much of the SAF would be used on international and intercontinental flights, and subsidies might prove too much for many jurisdictions.

It is useful at this point to review the cost estimates for SAF production, which have been described in the previous chapters. This is summarised in Table 6.3.

Table 6.3. Summary of Kerosene/Jet Fuel Production Costs

Method	Feedstock	Price	Jet fuel cost ($/bbl)	Relative to jet
Conventional jet fuel	Crude oil	$100/bbl	111	1.00
Ethanol	Fuel ethanol	$0.61/L	135	1.22
Olefins	Mixed olefins	$1,200/t	158	1.42
Vegetable oils	Carinata	$1,604/t	245	2.21
Biomass/methanol	Methanol	$490/t	241	2.17
Biomass/FTP	Biomass	$10/t	222	2.00
SAK by pyrolysis	Biomass	$10/t	99	0.89

With crude oil (Tapis) at $100/bbl, the conventional price of kerosene/ jet fuel is in the region of $111/bbl. Using ethanol as a base with an ethanol cost of $0.61/L, the production cost is about 10% higher at $135/bb. However, this cost is based on the traded costs of fuel ethanol for gasoline blending in the United States, which has the advantage of various grants and tax incentives to lower the differential with gasoline. Whether or not jet fuel would receive these incentives across all jurisdictions is a moot point. If fuel ethanol is used to make olefins, the production cost would be about $777/t, which is well below the traded price of olefins. If a more realistic price is used for olefins ($1,200/t), then jet fuel production cost would be about $158/bbl, nearly 50% higher than conventional petroleum jet fuel. This value can be taken as a guide to the production cost via ethanol without subsidies and tax incentives.

The production cost from vegetable oils (carinata) is estimated at $245/bbl, more than double the cost of petroleum jet fuel. This high cost is primarily a consequence of the high feedstock cost, which is taken to be similar to that of Canola, which is an immediate alternative for the farmer to grow and sell.

This case also illustrates a further point in that the alternative feedstocks contain significant quantities of oxygen. This has to be removed in one of two ways:

1. Rejection of carbon dioxide. This removes carbon from the feedstock and would increase the underlying cost of the carbon available for SAF production.

2. Hydrogenation of the oxygen to water. This will require significant quantities of hydrogen to deliver a truly sustainable SAF and should be made from relatively high-cost renewable sources.

These case studies emphasise the importance of feedstock costs. In order to minimise costs, biomass can be used as an alternative. Methanol via biomass costs is estimated to produce jet fuel for $241/t, which is slightly more than the Fischer–Tropsch route at $222/t. Again, it is more than double that for conventional jet fuel.

CONCLUSIONS

Status of SAF Technology

There are a variety of routes to sustainable aviation fuel (SAF) from sugars, vegetable oils and biomass. Sugars can be used to produce ethanol (or butanol), which can be easily converted into an olefin (known technology) and thence to a paraffinic kerosene (several known technologies).

The conversion of vegetable oils and fats to a fatty acid methyl ester (FAME) produces a fuel, which is suitable for a diesel substitute but is generally unsuitable for SAF. Vegetable oils and fats can be converted to paraffinic blendstock by commercially known hydrogenation and isomerisation. This produces a mixture of renewable (green) gasoline, kerosene and particularly diesel. Vegetable oil with high C_{22} and higher content can produce more jet-compatible blendstock. This process requires large quantities of hydrogen.

Biomass can be used to produce a paraffinic kerosene by gasification followed by the Fischer–Tropsch process or via methanol and olefins. Biomass can be used to produce a higher density kerosene blendstock by a series of hydrogenation–dehydrogenation processes or by pyrolysis followed by hydrogenation. These processes are under development.

These processes are, more or less, proven technologies for the production of SAF. Technologies for the production of SAF differ according to the feedstock being used. The main problem with the present known technologies is that the aviation fuel produced has too low a density for compliance with the current aviation fuel standards. This means that with the present specifications for jet fuel, the SAF is used in association with higher-density petroleum-based fuels.

Higher-density kerosene for blending can be produced from sustainable sources, but, at the time of writing, the technologies are largely unproven on a commercial scale.

Cost of SAF

Routes to SAF come at a production cost higher than conventional petroleum jet fuel. This is true for all viable routes. This will require airlines, and so on, to be willing to pay a premium for SAF. Since fuel cost is a major concern for airline operations, it is not clear if SAF will be viable except in niche operations.

The main reason for the higher cost of SAF is the underlying cost of the feedstock. Agricultural feedstocks are in competition with alternative uses for the feedstocks, for instance, food, biofuels for vehicles and chemicals. This places a floor price on the feedstocks available for SAF.

Waste agricultural products, such as sawdust, do not necessarily have alternative uses. However, these are generally in short supply and may offer lower feedstock costs for niche operations producing SAF.

Production of biomass on a larger scale places the biomass in competition with sustainable power energy-generating schemes, for instance, the production of wood chips for large-scale power generation in repurposed coal-fired generators. Again, this places a floor price on the biomass and increases the cost of the SAF.

APPENDICES

Relationship between the Cost of Feedstock and Petroleum Transport Fuels

Over the past few years, there has been increasing interest in the use of renewable feedstocks for the production of fuels. Each passing year sees more renewable feedstock grown to specifically satisfy the demand for renewable fuel. For example, a large portion of the world's Canola crop is utilised in the production of biodiesel as opposed to its use in food processing and as a cooking oil. As an example, 50% of Australia's annual export Canola crop of approximately 2.3 million tonnes goes to the production of biodiesel in Europe.[1]

It is often assumed that the cost of the feedstocks for renewable fuels is not linked to that of prevailing crude oil and its derivative motor fuels. The following data shows that this is not so and that there is a positive relationship between renewable feedstock and that of oil products. This is to be expected since the world consumption of crude oil (approximately 100 million barrels per day) represents a massive energy input into the world economy, and anything requiring energy would be expected to be subjected to the prevailing cost of crude oil. The extent of this relationship is illustrated in the following correlations.

---

[1] Australian Export Grains Innovation Centre, Brochure: Australian Canola, 2021.

Correlation of Canola Oil Price and Diesel (as Gas Oil) Price Post-2019[2]

The price of Canola oil is taken from the Canola Council of Canada for free on board (FOB) Vancouver; the Singapore spot price of gas oil is taken as reported by the OPEC.

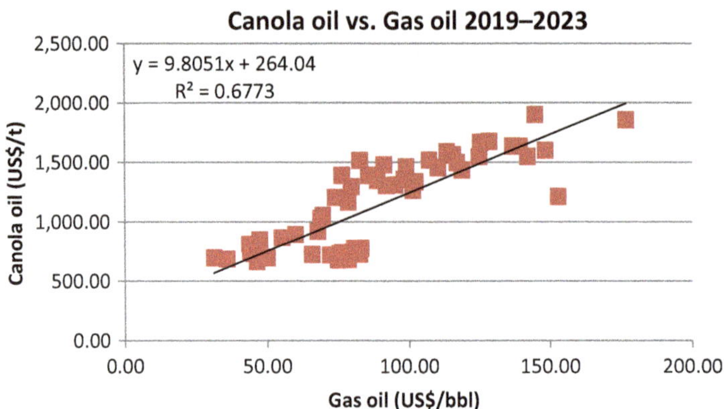

The positive correlation is quite strong with an *R*-squared factor of 0.68.

Correlation of Palm Oil Price and Diesel (as Gas Oil) Price 2019 to 2023

Palm oil prices are taken from the US Federal Reserve Economic Data, and gas oil from the OPEC data for the Singapore market.

For palm oil, although the correlation is positive, the correlation shows a much lower value of *R*-squared of 0.31 compared to that for Canola. This could be due to a larger number of uses for palm oil and derivatives in manufacturing, in particular, the large demand for soaps and the like.

---

[2] 2019 is chosen as the start-year representative of a time when there is an increasing use of vegetable oils in the production of motor fuels.

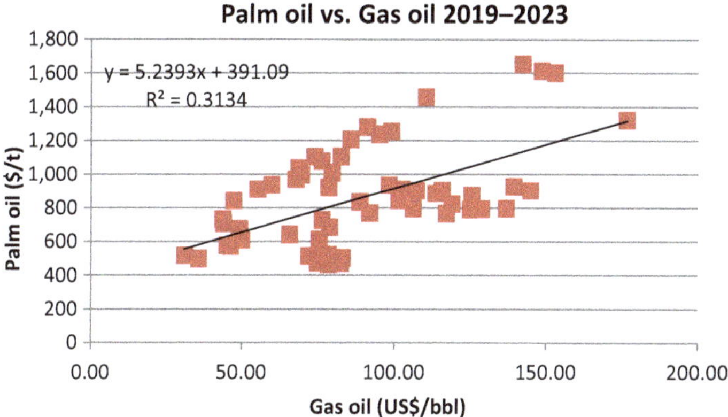

Correlation of Soybean Oil Price and Diesel (as Gas Oil) Price 2019 to 2023

Soybean oil prices are taken from the US Federal Reserve Economic Data and gas oil from the OPEC data for the Singapore market.

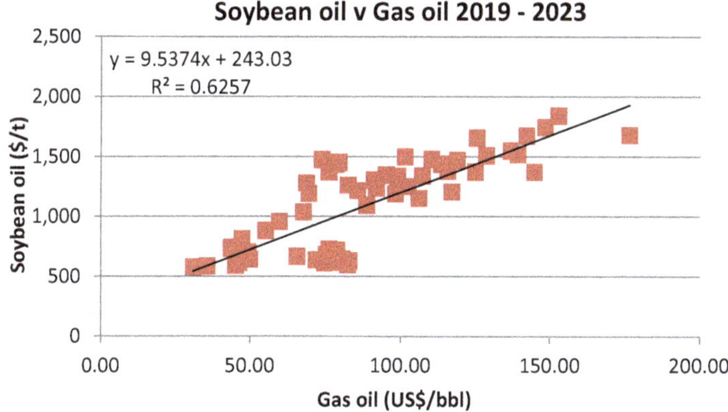

The correlation of soybean oil price with gas oil price is positive and quite good with an *R*-squared value of 0.63, similar to that for Canola.

Correlation of Ethanol Price and Premium Gasoline Price 2019 to 2023

Ethanol prices are taken from Trading Data and premium gasoline from the OPEC data for the Singapore market.

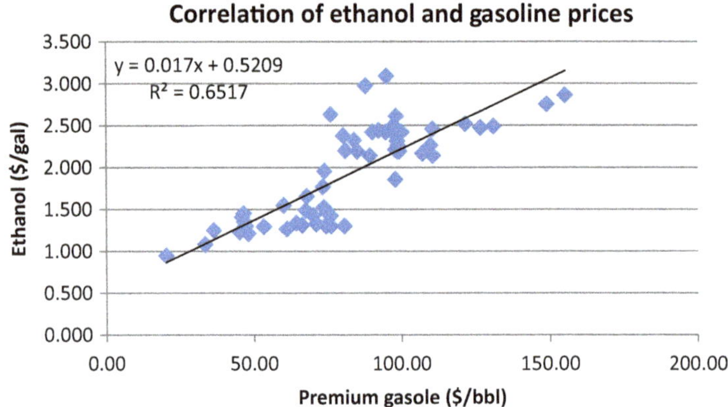

The correlation for ethanol and premium gasoline shows a positive correlation with an R-squared value of 0.65.

Conversion Factors

Conversion factors		
Cubic metre	35.315	Cubic feet
m^3@15°C	35.383	Ft^3@60°F
GJ	0.9478	MMBTU
$/GJ	1.055	$/MMBTU
1 kWh	3.6	MJ
lb	0.4536	kg
HP	0.7457	kW
GWh	3.6E-03	PJ
US Gallon	3.78541	Litre

(Continued)

Temperature		
Absolute zero	C	−273.15
Normal	C	15
Gas constant	R	1.9859
	J/mol/K	8.3145
Days per year		340
Hours per year		8160
Nautical miles	1.852	km

Note: M = Mega (million in SI); MM = million
US customary units

Methodology for Economic Analysis

What is required is a rapid approach to determine the economic viability of a particular technology of interest, namely, a concept analysis where speed is not gained at the expense of accuracy. This requires a systematic approach in which various technologies and approaches are treated in the same manner so that the economics from one route to sustainable aviation fuel (SAF) can be compared to another.

The methodology described was devised by the ICI PLC in order to evaluate all of the diverse routes to the production of ethylene from any feedstock using widely disparate technologies with different plant construction periods and years of operation. The methodology has been published by Stratton et al.[3] and is generally applicable for energy-intensive industries. The basic economic equation is

$$P = F + C + O,$$

where P, the unit production cost of the product of interest (SAF say), is equal to the sum of the unit feedstock costs (F), the unit capital costs (C) and the unit non feedstock operating costs (O). This can be expressed as a fixed-variable equation with the fixed part of the equation representing

---

[3]A. Stratton, *A Simplified Method of Calculating product Cost*, Technical Note 3, Economic Assessment Service, IEA Coal Research, 1982.

the return on capital (the unit capital costs, C, independent of tax considerations) together with all the unit non-feedstock operating costs (O).

Capital Costs (C)[4]

The capital costs are developed for greenfield projects completely isolated from other facilities. All the costs associated with utilities (unless otherwise accounted) are allowed for in the capital cost. Some processes require small amounts of power. This is considered an import.

Capital is estimated using the published information and using the location factors, and Nelson–Farrer Indices are adjusted to the US Gulf site and 2022 costs for all processes. Scaling used the exponent method, namely,

Capital of Plant [1]/Capital of Plant [2] = {Capacity of Plant [1]/ Capacity of Plant [2]}^{n},

where n is a constant with a value that is typically 0.70 to 0.75.

The capital cost of facilities in other locations can be estimated from the factors given here.

Capital Recovery Factors

For a plant with a capital cost of Co, the plant investment cost, C, capitalises the return on investment during construction of the plant, that is takes account of expenditure and the required return during the construction period.

$$C = Co \sum_{s=0}^{p} a_s (1+i)^{p-s},$$

where a_s represents the breakdown of capital expended over the construction period, p is the first year of production, and s is a general year of the

---

[4] See also T.R. Brown, Capital Cost Estimating, *Hydrocarbon Processing*, Gulf, October, 2000, p. 93 for a discussion of various cost estimating methods and J.T. Summerfeld, Ibid., June, 2001, p. 103 for an analysis of estimation accuracy.

project starting $s = 0$, with construction complete at $s = p$. The return on investment is i. The values of a_s used in this work are given in the table here:

Values for $a(s)$

Construction period	1 year	2 years	3 years	4 years	5 years
Values of $a(s)$					
$a(0)$	100%	50%	30%	17%	4%
$a(1)$		50%	45%	32%	14%
$a(2)$			25%	26%	32%
$a(3)$				25%	36%
$a(4)$					14%

The general DCF equation can be written:

$$C = \sum_{r=1}^{N} \left(\mathrm{Rr} - \mathrm{FCr} - \mathrm{Vcr} \right) \big/ \left(1 + i\right)^{r},$$

where r is the production year, with N the final production year and Rr is the total product revenue in year r, FCr is the fixed costs in year r and VCr is the variable cost in year r.

This equation is simplified by assuming that there is no build-up to full production, and full production is achieved as soon as construction is complete. This is followed by N years of full production. Hence:

$$C\left(1 + i\right) = \left(\mathrm{Rr} - \mathrm{FCr} - \mathrm{Vcr} \right) \sum_{r=1}^{N} 1 \big/ \left(1 + i\right)^{r}.$$

This is rearranged to give:

$$\mathrm{Rr} = \mathrm{FCr} + \mathrm{VCr} + K(1 + i)\, C,$$

where K is the sum of the geometric series:

$$K = i\,(1 + i)N/[(1 + i)N - 1].$$

The values of K for various values of i (required rate of return) and N (operating life) are given in the table here:

Values for K

N	10	15	20	25	30
Interest (i)					
5.00%	0.1295	0.0963	0.0802	0.071	0.0651
7.50%	0.1457	0.1133	0.0981	0.0897	0.0847
10.00%	0.1627	0.1315	0.1175	0.1102	0.1061
12.50%	0.1806	0.1508	0.1381	0.1319	0.1288

The capital recovery factor (K_o) is then:

$$K_o = K (1 + i) C,$$

and from the table, we get:

$$Ko = K(1+i)\, Co \sum_{s=0}^{p} a_s (1+i)^{p-s}.$$

The values for the return on capital (ROC) or Ko/Co are given in Tables A7-3 and A7-4 for a royalty-free basis and one encompassing a 2% royalty to the process licensor, respectively. The first table has been used

Values for Return on Capital (Royalty-Free Basis)

1-Year construction						
	N	10	15	20	25	30
	Interest (i)					
	5.00%	13.60%	10.12%	8.43%	7.45%	6.83%
	7.50%	15.66%	12.18%	10.54%	9.64%	9.10%
	10.00%	17.90%	14.46%	12.92%	12.12%	11.67%
	12.50%	20.32%	16.96%	15.54%	14.84%	14.49%
	15.00%	22.91%	19.67%	18.37%	17.79%	17.51%
2-Year construction						
	N	10	15	20	25	30
	Interest (i)					
	5.00%	13.94%	10.37%	8.64%	7.64%	7.00%

(*Continued*)

		10	15	20	25	30
	7.50%	16.25%	12.64%	10.94%	10.01%	9.44%
	10.00%	18.80%	**15.19%**	**13.57%**	12.72%	12.25%
	12.50%	21.59%	18.02%	16.51%	15.77%	15.39%
	15.00%	24.63%	21.14%	19.75%	19.12%	18.83%
3-Year construction						
	N	**10**	**15**	**20**	**25**	**30**
	Interest (i)					
	5.00%	14.32%	10.65%	8.87%	7.85%	7.19%
	7.50%	16.92%	13.16%	11.39%	10.42%	9.83%
	10.00%	19.84%	**16.02%**	**14.32%**	13.43%	12.93%
	12.50%	23.08%	19.27%	17.65%	16.86%	16.45%
	15.00%	26.68%	22.90%	21.39%	20.71%	20.39%
4-Year construction						
	N	**10**	**15**	**20**	**25**	**30**
	Interest (i)					
	5.00%	14.59%	10.85%	9.04%	7.99%	7.33%
	7.50%	17.39%	13.52%	11.71%	10.71%	10.11%
	10.00%	20.58%	16.62%	14.85%	13.93%	13.41%
	12.50%	24.17%	20.18%	18.48%	17.66%	17.23%
	15.00%	28.20%	24.21%	22.61%	21.90%	21.56%
5-Year construction						
	N	**10**	**15**	**20**	**25**	**30**
	Interest (i)					
	5.00%	14.71%	10.94%	9.11%	8.06%	7.39%
	7.50%	17.61%	13.69%	11.85%	10.84%	10.23%
	10.00%	20.91%	16.89%	15.09%	14.16%	13.63%
	12.50%	24.66%	20.58%	18.85%	18.01%	17.58%
	15.00%	28.87%	24.78%	23.15%	22.42%	22.07%

Values for Return on Capital with 2% Royalty

1-Year construction						
	N	**10**	**15**	**20**	**25**	**30**
	Interest (i)					
	5.00%	13.87%	10.32%	8.59%	7.60%	6.97%
	7.50%	15.97%	12.42%	10.76%	9.84%	9.28%
	10.00%	18.26%	14.75%	13.18%	12.36%	11.90%
	12.50%	20.73%	17.30%	15.85%	15.14%	14.78%
	15.00%	23.37%	20.06%	18.74%	18.15%	17.86%
2-Year construction						
	N	**10**	**15**	**20**	**25**	**30**
	Interest (i)					
	5.00%	14.22%	10.58%	8.81%	7.79%	7.14%
	7.50%	16.57%	12.89%	11.16%	10.21%	9.63%
	10.00%	19.17%	**15.49%**	**13.84%**	12.98%	12.50%
	12.50%	22.02%	18.38%	16.84%	16.09%	15.70%
	15.00%	25.13%	21.56%	20.15%	19.51%	19.20%
3-Year construction						
	N	**10**	**15**	**20**	**25**	**30**
	Interest (i)					
	5.00%	14.61%	10.87%	9.05%	8.00%	7.34%
	7.50%	17.26%	13.42%	11.62%	10.63%	10.03%
	10.00%	20.23%	**16.34%**	**14.60%**	13.70%	13.19%
	12.50%	23.54%	19.65%	18.00%	17.20%	16.78%
	15.00%	27.21%	23.36%	21.82%	21.13%	20.80%
4-Year construction						
	N	**10**	**15**	**20**	**25**	**30**
	Interest (i)					
	5.00%	14.88%	11.07%	9.22%	8.15%	7.47%
	7.50%	17.74%	13.79%	11.94%	10.92%	10.31%
	10.00%	20.99%	16.96%	15.15%	14.21%	13.68%

(Continued)

		10	15	20	25	30
	12.50%	24.66%	20.58%	18.85%	18.01%	17.58%
	15.00%	28.77%	24.69%	23.06%	22.33%	21.99%
5-Year construction						
	N	**10**	**15**	**20**	**25**	**30**
	Interest (i)					
	5.00%	15.00%	11.16%	9.29%	8.22%	7.53%
	7.50%	17.96%	13.96%	12.09%	11.06%	10.44%
	10.00%	21.33%	17.23%	15.39%	14.44%	13.90%
	12.50%	25.15%	20.99%	19.23%	18.37%	17.93%
	15.00%	29.45%	25.28%	23.61%	22.87%	22.51%

for typical non-process items (pipelines, ships etc.), and the second table has been used for licensed processes.

The selection of a rate of capital return is dependent on many factors, including the nature of the industry in question. For upstream oil and gas developments, or relatively small-scale process plant, the high rates of capital return are often demanded by the investors to offset short operational lives or perceived higher levels of risk. For very long-term (30 years) infrastructure projects often accessing government funds, the far lower rates of return are required. Many Greenfield operations in the process industries are planned for a lifetime of 15 to 20 years, and the rates of return are as appropriate. The commonly used values for the return on capital in this work are emboldened in the tables.

Fixed Operating Costs (O)

Working Capital. Rather than capitalise the working capital and handling it with the project capital (Stratton), the working capital is treated as an annual operating cost. The reasoning behind this is that working capital is normally borrowed against the business and is fully recovered at the end of the project. The outgoings are the interest on the debt. The value of working capital can be taken as 5% of the plant capital or 30 days stock. The latter is generally smaller than the former and was used when sufficient data permitted its calculation.

Labour, Maintenance and Administrative Costs. As a general rule, labour and maintenance were each charged at the rate of 3% of the capital per annum. For labour, this included both direct and indirect labour costs. For maintenance, this included both materials and labour. Over the past decades, many companies have made attempts to reduce the operating labour and maintenance charges. Labour can be reduced by extensive computer control. However, the success in reducing the maintenance charge is difficult to quantify, and several operations have suffered major problems claimed to be due to the cutbacks in maintenance costs. Administrative costs are basically insurance and local land taxes. A value of 1.5% of the fixed capital as an annual charge was used.

Catalysts and Chemicals. Most plants require some chemicals for water treatment purposes. Catalyst charges are based on a 3- to 5-year turnaround.

Other Operating Costs. Some processes require inputs other than the principal hydrocarbon feed. This is usually electric power, and the typical average values were used.

Location Factors[5]

Capital (CAPEX) and Operating Cost (OPEX) Factors

	1	2	3	4
Climate/terrain	Benign	Difficult	Difficult	Extreme
Gas transmission	Present	Present	No	No
Freshwater	Present	Present	No	No
Ship loading	Present	Present	No	No
Employee housing	Present	Present	No	No
Labour costs	Low	High	High	High
Relative CAPEX	1.000	1.155	1.562	2.250
Relative OPEX	1.000	1.139	1.520	2.039

---

[5] Assessment of Cost Benefits of Flexible and Alternative Fuel Use in the US Transportation Sector — Technical Report Three — Methanol Production and Transportation Costs, United States Department of Energy, November 1989.

(*Continued*)

	1	**2**	**3**	**4**
Examples	US Gulf	Urban Australia	Remote FE	Offshore
	Canada	New Zealand	Remote Aus	Arctic
		Developed FE		
		Middle East		

Relation of Delivered Equipment to Final Capital Cost[6]

	Solids	**Solid–Fluid**	**Fluid**
Equipment	100%	100%	100%
Installation	45%	39%	37%
Piping	16%	31%	93%
Structural foundations			7%
Electrical	10%	10%	10%
Instruments	9%	13%	
Building services	25%	39%	26%
Site preparation	13%	10%	10%
Auxiliaries	40%	55%	70%
Subtotal plant	258%	297%	353%
Field expense	39%	34%	41%
Engineering	33%	32%	33%
Direct plant costs	330%	363%	427%
Contractor fees	17%	18%	21%
Contingency	34%	36%	42%
Capital investment	381%	417%	490%

Note that the equipment cost is about one-quarter of the final cost. Often, when a new process or product comes to market, the initial cost is very high. As time progresses, manufacturers of the product continually improve production techniques and introduce improvements in the

---

[6]R.H. Perry and D.W. Green, *Perry's Chemical Engineers' Handbook,* 7th ed. McGraw-Hill, 1997: Table T 9-51.

efficiency of production. Production efficiencies are also improved by the scale of production — the more gadgets produced lower the fixed cost of the production. The result is that the cost of the product falls. This can be quite dramatic in the early days of the introduction of a new product. This phenomenon is well known and often referred to as a learning curve.

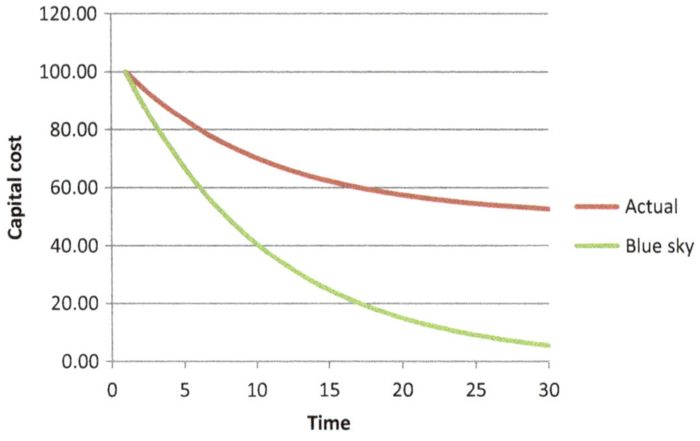

The learning curve

However, there are limits to the extent of this fall, which over time levels off and reaches an asymptote. All other things being equal, the level of the asymptote (lowest production cost) is determined by:

- The cost of labour to assemble the item.
- The cost of energy used to produce the item.
- The cost of other material in the construction of the item.

In writing this book, the author has come across several articles that appear to assume a constant fall in the capital cost of a specific item so that, at some future date, a project becomes profitable. This particularly concerns the cost of renewable technology. For example, the cost of power from PV systems, where 10-year projections of cost predict very low power generation cost. It is the view of the author that this is a blue sky scenario as it assumes the fall in the cost of the solar voltaic cell is replicated in the cost structure of the system, which is only a small part of the finished system.

Specific Volume and Heating Values for Fuels

	L/t	bbl/t	HHV GJ/t	LHV GJ/t
Acetylene			49.92	48.17
Ammonia			22.50	18.60
AVTUR	1,261	7.93	46.40	
n-Butane			49.50	45.73
iso-Butane			49.41	45.61
Butanes	1,928	12.13	49.45	45.67
But-1-ene			48.45	45.31
Butenes			48.10	45.00
Carbon			32.80	32.80
Carbon monoxide			10.09	10.09
Crude oil (35 API)	1,177	7.40	45.00	
Crude oil (40 API)	1,212	7.62		
Diesel	1,182	7.43	45.90	43.00
DME	1,493		31.00	28.40
Ethane	2,654		51.88	47.49
Ethylene			50.29	47.15
Fuel oil (LS)	1,110	6.98	44.10	
Gasoline	1,360	8.56	46.70	42.50
Hydrogen			141.86	119.93
LPG		12.35	49.87	46.01
Methanol	1,272	8.00	22.70	19.50
Methane			55.56	50.07
Natural gas			53.90	
Naphtha	1,534	9.00	48.10	
Propane	1,998	12.57	50.33	46.36
Propylene			48.95	45.77
LNG	2,197.80	13.83	54.4	
TAPIS crude	1,231.07	7.75	43.52	

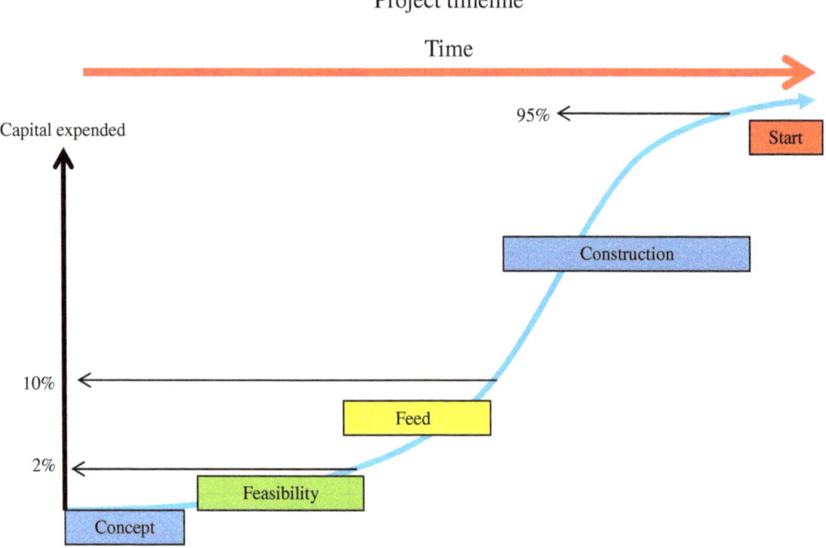

Figure A.1. Project Timeline and Different Levels of Cost Estimation

Accuracy of Cost Estimates

The cost estimates fall into four types: concept, feasibility, bankable feasibility and front end-engineering and design. These types of study are performed at different stages of a project as illustrated in Figure A.1.

This book concerns the first concept stage from which an optimum concept can be defined and the key issues for project feasibility identified.

Concept or Scoping Study

OBJECTIVE: To generally define the project,

- define the product and by-products
- identify feedstock and price range
- identify the product market and sales price range

- identify the project scale
- identify the technology to be employed
- identify provisional mass and energy balance for process
- identify process licensors/EPC contractors
- identify potential location
- identify order estimate for CAPEX and OPEX
- identify viability of the concept

STUDY LOCATION: Home office
COST: <$100,000.
ERROR in cost estimate: ±25% to 40%.

Feasibility Study

OBJECTIVE: To define a specific project,

- define the product and by-products in terms of purity, and so on
- define the product market and outline off-take agreements
- identify feedstock source, quality and supplier and obtain outline price agreements
- identify **E**ngineering, **P**rocurement and **C**onstruction contractor appointed
- define the project location and scale
- identify the technology to be employed and obtain supplier details
- define mass and energy balance of all major unit operations
- obtain budget cost estimates for major items
- size major units, identify constructors and define installation defined
- define and cost utilities and off-sites
- determine estimate for CAPEX and OPEX

STUDY LOCATION: Home office and EPC contractor office.
COST: 1% to 2% of final CAPEX.
ERROR in cost estimate: ±15%.

Bankable Feasibility Study

OBJECTIVE: To help in raising capital from third parties,
     All of the actions for a feasibility study plus:

- written and executed feedstock off-take agreements
- written and executed product and by-product off-take agreements
- environmental impact statement — approvals commence
- review of feasibility study by independent EPC contractor
- review of project feasibility by independent auditor
- production of *pro-forma* revenue and balance sheet
- production of *product disclosure statement*
- preliminary discussions with potential investors

STUDY LOCATION: Home office/Investment advisor office.
COST: 3% or more of final CAPEX; can be high for large projects.
ERROR in cost estimate: ±15%!

Front-End Engineering and Design

OBJECTIVE: To finalise CAPEX and OPEX before the final investment
decision,

- All government approvals
- EPC contractor produces engineering design for all major items of equipment
- EPC contractor identifies manufactures of equipment
- EPC contractor obtains cost estimates for equipment procurement
- EPC contractor defines and engineers major utilities and off-sites
- EPC contractor defines construction logistics
- EPC contractor produces a final estimate for CAPEX and OPEX

STUDY LOCATION: EPC contractor home office.
COST: 10% or more of final CAPEX; can be higher for large projects.
ERROR in cost estimate: ±5%.

THE ERROR IN THE COST ESTIMATE IS INVERSELY PROPORTIONAL
TO THE MONEY SPENT.

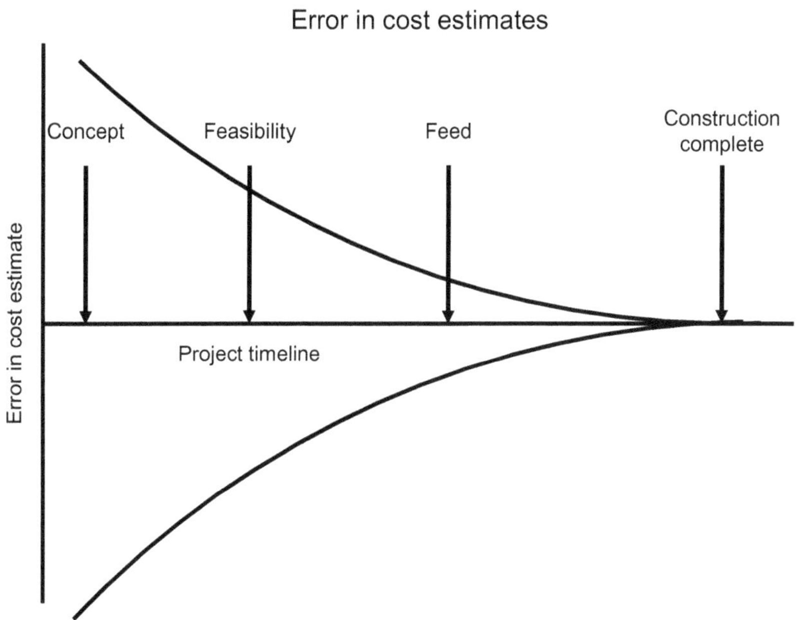

Exotic Biofuel Feedstocks

Exotic feedstocks are described in the following sheets. Where available and to help identification, a picture of the plant, flower or nut is provided.[7] To help avoid confusion, a scientific name is provided. Sometimes, a particular plant has been categorized differently, and there may be two or more scientific descriptions for the same plant.

A table of typical fatty acid composition for the most common fatty acids is provided. The total is the sum of these common fatty acids. Values below 100% can be taken as a guide for the amounts of other fatty acids

---

[7] The photographs were downloaded from Wikipedia. Reference is given to the author of the picture where details of the copyright are available. No changes have been made to the pictures.

present. Where there is a large discrepancy (e.g. coconut and kapok), further comment is made. Total values over 100% can be taken as errors in the analysis. As noted earlier, there is a large geographic and seasonal variation in the analysis of plant fatty acid compositions, so values in the tables should be taken as a guide only.

A table of general fatty acid methyl ester (FAME) properties and biodiesel properties, which are dependent upon the fatty acid composition is presented. The first row of the table is the estimates based on the fatty acid composition, including minor fatty acid components. The estimates presented are for the Saponification Number (S.N.), Iodine Number (I.N.), Cetane Number (C.N.), Density (kg/m^3), Viscosity (cSt) and the % Distilled at 360°C (%D(360)). The methods of estimating have been discussed earlier. The second row gives pertinent literature values for the FAME. Note that due to the paucity of comprehensive analysis, most of the literature data is from different sources to the fatty acid composition. Sometimes, no relevant literature values are available.

Issues relating to the use of the FAME as a biodiesel fuel are discussed. This generally takes the form of a comparison to the principal feedstocks (Canola, soybean, rapeseed, palm olein and tallow), commenting on similarities and differences.

Finally, brief relevant agricultural and cultivation issues are noted.

Algae

Fatty Acid Composition[8]

	C16:0	C18:0	C18:1	C18:2	C18:3	C22:1	Total
Algae (1)	6.7	3.3	33.9	54.5	0.4		98.8
Algae (2)	5.0	1.3	63.2	21.4	7.1		98.0

Algae (1) *Nannochloropsis;* Algae (2) *Chlorella*

Lipid composition of algae oil varies according to species and growing conditions. There seems to be a wide variation.

---

[8]A. Kleinova, Z. Cvengrosova, J. Rimarcik, E. Buzetzki, J. Mikulec and J. Cvengros, Biofuels from Algae, *Proc. Eng.* **42**, 232–238 (2012).

Biodiesel Properties

	S.N.	**I.N.**	**C.N.**	**Density**	**Viscosity**	**%D(360)**
Estimate	202.1	121.4	46	865	3.02	98.8
Literature						

Biodiesel Issues

On the face of it, these values for algae FAME would produce a good-quality biodiesel fuel. However, there is a paucity of data on the FAME produced from the various forms of algae. Some analysis indicates the presence of very high molecular weight and highly unsaturated oils (similar to fish oils, see here). Conventional wisdom is that these will be discarded during work-up and production of the biodiesel.

Agricultural Issues

For the most part, we are not concerned with the production of algae in ponds, rather a semi-industrial mass production of algae in large plastic tubes through which the algae is continually being circulated in a carbon dioxide/water reticulation system exposed to sunlight. The development of these systems, including the selection of the particular genus of algae for optimum biodiesel production is linked with the potential for carbon dioxide sequestration from fossil fuel-emitting processes. There are several research and demonstration projects underway.

The production of hydrocarbon oil from *Botryococcus braunii* has the composition[9]:

Isobotryococcene	4%
Botrycoccene	9%
$C_{34}H_{58}$	11%
$C_{36}H_{62}$ (isomer A)	34%
$C_{36}H_{62}$ (isomer B)	4%
$C_{37}H_{64}$	20%
Other hydrocarbons	18%

---

[9]Wikipedia, downloaded August 15, 2024.

Candlenut[10]

Scientific Name: *Aleurites moluccana*
Other Names: Kukui, see also Tung

Fatty Acid Composition

	C16:0	**C18:0**	**C18:1**	**C18:2**	**C18:3**	**C22:1**	**Total**
Candlenut	5.5	6.7	10.5	48.8	28.5	0	100

Estimated Biodiesel Properties

	S.N.	**I.N.**	**C.N.**	**Density**	**Viscosity**	**%D(360)**
Estimate	201.1	176.1	33.8	885	3.71	100
Literature		164.3				

Biodiesel Issues

Broadly similar to soybean but with higher levels of the C18:3 fatty acid, this will impart a higher level of instability on the product biodiesel. The high level of reactive olefins is confirmed by the high iodine number of

---

[10]Wikipedia, *Picture Author Forest and Kim Starr*, downloaded July 22, 2024.

over 160. This will require a higher level of additives to suppress polymerisation and oxidation.

Candlenut is a potential biodiesel but not in its own right, rather blended with more saturated feedstocks such as tallow, which will lower the final level of highly unsaturated C18:3 species.

Agricultural Issues

The candlenut grows to a height of 30 m with wide-spreading and pendulous branches. The tree is now widespread in the tropics, particularly in forests. It is fast growing. It has been harvested and cultivated for millennia. The fruit is a drupe (4–6 cm in diameter) with one or two lobes. Each lobe has a single soft, white oily, kernel contained within a hard shell that is about 2 cm in diameter. The fruit is mildly toxic when eaten raw. The kernel is the source of the candlenut oil.

Canola[11]

Scientific Name: *Brassica napus*

---

[11] Wikipedia, *Picture Author — Canada Hky*, downloaded July 22, 2024.

Fatty Acid Composition

	C16:0	C18:0	C18:1	C18:2	C18:3	C22:1	Total
Canola	3.9	3.1	60.2	21.1	11.1	0.5	99.9

Estimated Biodiesel Properties

	S.N.	I.N.	C.N.	Density	Viscosity	%D(360)
Estimate	199.8	123.3	45.9	881	4.16	99.5
Literature		115	59	885	4.66	100

Biodiesel Issues

Canola is one of the principal feedstocks for biodiesel. The literature values are for a low (0%) erucic acid rape methyl ester. Canola is a variant of rapeseed with low erucic (C22:1) acid content. It has been developed and used in Canada as a cooking oil and as feedstock for biodiesel. Internet sources indicate that a significant amount of the rape grown in Europe for biodiesel production is in fact Canola sometimes referred to as low erucic acid rapeseed.

Agricultural Issues

Canola is now a generic term for edible varieties of rapeseed oil in North America and Australasia. It is a major agricultural crop for cooking oil and cattle feed. Genetically modified versions are becoming increasingly used, for example, for herbicide tolerance.

The 2019 production of Canola was 24Mt. There is a major international trade in Canola dominated by Canada (over 3Mt/y).

Coconut[12]

Scientific Name: *Cocos nucifera*

Fatty Acid Composition

	C16:0	C18:0	C18:1	C18:2	C18:3	C22:1	Total
Coconut	9.2	2.9	6.9	1.7	0	0	20.7

In coconut, most of the fatty acids are lighter:

C8:0	C10:0	C12:0	C14:0	C16:0	C18:0	C18:1	C18:2	Total
8.3	6	46.7	18.3	9.2	2.9	6.9	1.7	100

Estimated Biodiesel Properties

	S.N.	I.N.	C.N.	Density	Viscosity	%D(360)
Estimate	270.5	9.2	64.4	871	2.55	100
Literature		8.5	58.8	875	2.75	100

---

[12] Wikipedia, *Picture Author Judgefloro*, downloaded July 23, 2024.

Biodiesel Issues

Coconut oil has a high level of saturation, which imparts good storage and oxidative stability to the biodiesel it produces. The problem with coconut as a biodiesel feedstock, which has been widely examined, is the high volume of C8, C10 and C12 fatty acids. As esters, these lower the boiling point of the product. For example, C8 methyl ester has a boiling point of 193°C, which is at the top of the gasoline range and well below the typical initial boiling point of automotive diesel. The reported flash point is 103°C is below the biodiesel standard value.

However, the low molecular weight of the fatty acids makes it an excellent feedstock for bio-jet fuel production. Bio-jet does not use FAME rather is produced by fully hydrogenating the oil triglycerides to give hydrocarbons boiling in the C10 to C14 range.

Agricultural Issues

It is a well-known agricultural crop throughout the tropics. The coconut is not a true nut. A palm (up to 30 m tall) produces about 75 fruits/y. The fruit weighs about 1.4 kg. The oil is derived from the kernels, meat and milk of the coconut. There are several processes for extracting the oil, one of which involves extraction with hexane.

Copaiba[13]

---

[13] Wikipedia, *Picture Author Mauroguanandi*, downloaded July 23, 2024.

Scientific Name: *Copaifera officinalis, langsdorffii and reticulata*
Other Names: Diesel tree

Fatty Acid Composition

The analysis for *C. langsdorffii* is given below:

	C16:0	**C18:0**	**C18:1**	**C18:2**	**C18:3**	**C22:1**	**Total**
Copaiba	12.71	4.36	30.96	45.34	0	0	93.37

Estimated Biodiesel Properties

	S.N.	**I.N.**	**C.N.**	**Density**	**Viscosity**	**%D(360)**
Estimate	199.6	111.2	48.6	879	3.64	93.4
Literature						

Biodiesel Issues

The tree has been researched for the production of diesel by tree-tapping. From the above analysis of the fatty ester distribution, biodiesel could be potentially produced from the seed oils.

Agricultural Issues

The tropical rainforest tree *C. langsdorffii* is known as the diesel tree and kerosene tree. It produces a large amount of terpene hydrocarbons in its wood and leaves. One tree can produce 30 to 40 L of hydrocarbons per year. The oil is collected by tree tapping. The main compound in the oil is copaiba, an oleoresin that is useful in the production of oil products such as lacquers and can be used as biodiesel. The tree is also the main source of copaene, another terpene.

It is a medium-sized tree usually reaching 12 m in height, with white flowers and small, oily fruits. The wood is light due to its porosity. It is honeycombed with capillaries filled with oil. Tapping the tree involves cutting a well into which the oil seeps and where it can be easily collected. Despite its vigorous production of oil, the tree does not grow well outside of the tropics and does not show promise as a reliable source of biodiesel.

Presumably, the last reference is for biodiesel produced from tapping the tree rather than the nuts.

Corn (*Maize*)[14]

Scientific Name: *Zea mays*
Other Names: Maize

Fatty Acid Composition

	C16:0	C18:0	C18:1	C18:2	C18:3	C22:1	Total
Corn	22.9	3.1	18.5	54.2	0.5	0	99.2

Estimated Biodiesel Properties

	S.N.	I.N.	C.N.	Density	Viscosity	%D(360)
Estimate	204.3	116.4	46.8	884	4.00	100
Literature		120.5				

---

[14] Wikipedia, *Picture Author Christian Fischer*, downloaded July 23, 2024.

Biodiesel Issues

Corn oil has a very similar composition to soybean oil and would be expected to have similar properties to soybean biodiesel. The level of saturation is higher than soybean and may have better oxidative resistance than soybean.

Agricultural Issues

It is a major world food crop and grown extensively throughout the world. In 2020, the world production was 1.1 billion tonnes. There are a large number of cultivars. Corn is extremely versatile although it prefers hot, wet climates; it can also thrive in cold, hot, dry or wet conditions. In the United States, large amounts are grown for the production of ethanol for fuel use.

Cottonseed (Cotton)[15]

Scientific Name: *Gossypium hirsutum (American); G. barbadense (Egyptian); G. herbaceum (Asiatic)*

Fatty Acid Composition

	C16:0	C18:0	C18:1	C18:2	C18:3	C22:1	Total
Cotton	23.9	2.7	18.2	52.45	0.2	0.15	97.6

[15] Wikipedia, *Picture Author US Department of Agriculture*, downloaded July 23, 2024.

As well as the principal fatty acids of interest, cottonseed oil contains some minor quantities (2%) of higher unsaturated fatty acids. Detailed analysis of these acids can be used to detect its presence in food quality oils, for which it is generally not recommended. Of more importance is the presence of cyclopropenoid fatty acids which impart a toxic nature to the oil.

Estimated Biodiesel Properties

	S.N.	I.N.	C.N.	Density	Viscosity	%D(360)
Estimate	204.2	112.9	47.6	885	3.88	99
Literature		109				

Biodiesel Issues

As a biodiesel feedstock, the fatty acid distribution resembles that of soybean. The minor quantities of unsaturated molecules and the cyclopropenoid compounds should not unduly cause a problem. However, these may lead to deterioration in the relative oxidative stability of the fuel. There should be no issues with its use as a blend component with more stable palm olein or tallow.

Agricultural Issues

Successful cultivation requires a long frost-free period, plenty of sunshine and moderate rainfall. The 2022 production was 69.7Mt. Cotton is a widely grown crop for fibre and clothing manufacture. There is a large international trade. The large volume of cotton production produces a significant amount of cotton seed by-product.

The world production of cottonseed oil is about 4.5Mt/y.

Crambe[16]

Scientific Name: *Crambe Abyssinica*

Crambe is a genus of *Brassicaceae* native to Europe, southwest and central Asia and eastern Africa. There are about 20 species: for example, *Crambe maritima; C. cordifolia; C. tataria*

Fatty Acid Composition

	C16:0	C18:0	C18:1	C18:2	C18:3	C22:1	Total
Crambe	1.7	1	16.7	7.8	6.9	55.7	89.8

This analysis relates to *C. abyssinica*. The full analysis indicates the presence of higher molecular weight fatty acids.

Estimated Biodiesel Properties

	S.N.	I.N.	C.N.	Density	Viscosity	%D(360)
Estimate	180.2	97.7	54.6	876	4.67	35.5
Literature		93.0				

---

[16] Wikipedia, *Picture Author Kurt Stuber*, downloaded July 23, 2024.

Biodiesel Issues

Crambe has a very high value of C22:1 fatty acid (erucic acid). This makes it very similar to rapeseed oil. The biodiesel would not meet the distillation (T90) standard. The high C22:1 content would make it attractive for hydrocracking to produce C11 hydrocarbons for use in SAF.

Agricultural Issues

This is a winter crop in southern Europe and a spring crop in northern Europe. It is considered drought tolerant.

There are several cultivars and the habit of all is not known. They are grown in gardens as a decorative border plant. The crop is grown in the United States for its erucic acid component, which before the advent of Canola was sourced from rapeseed.

False Flax[17]/*Camelina sativa*

Scientific Name: *Camelina sativa*
Other Names: Gold-of-pleasure, wild flax, linseed dodder, camelina, German sesame, and Siberian oilseed

---

[17]Wikipedia, *Picture Author Unknown*, downloaded July 24, 2024.

Fatty Acid Composition

	C16:0	C18:0	C18:1	C18:2	C18:3	C22:1	Total
False Flax	5.5	2.5	19.5	15.5	36	1	80

There is a significant amount (16%) of unsaturated C20:1 and other C20 unsaturated acids.

Estimated Biodiesel Properties

	S.N.	I.N.	C.N.	Density	Viscosity	%D(360)
Estimate	191.6	161.6	38.4	885	3.37	79.0
Literature		157.5			4.38	

Biodiesel Issues

The fatty acids have a very high level of unsaturation as evidenced by the high iodine number and C18:3 content. The oil itself is very stable due to a high content of natural antioxidants (tocopherols). However, produced biodiesel (FAME) would be oxidatively unstable. There would also be an issue with the cetane number and in achieving the T(90) distillation point. False flax is probably not suitable as a biodiesel feedstock. However, this source has been used in part to produce synthetic kerosene, which has been blended with conventional petroleum fuel to produce an SAF.

Agricultural Issues

This is a summer or winter annual crop and is cultivated across the world to produce vegetable oil and animal feed. It is widespread in North America. Probably, it has a tendency to become a weed.

Fish Oil/Marine Oils

Fish oils and marine oils are an important group of oils that are harvested for their high levels of high molecular weight, highly unsaturated fatty acids (polyunsaturated *w*-3 and *w*-6 fatty acids).

Fatty Acid composition

There is a wide variation depending on species type and food source.

	C16:0	C18:0	C18:1	C18:2	C18:3	C22:1	Total
Red sea bream	22.5	6.2	17.3	4.4	0.9	0.0	51.3
Tuna	17.6	4.1	14.8	1.9	0.0	0.5	38.9
Whale (pilot)	8.3	2.0	51.1	1.4	0.0	0.6	63.4

Other significant acids present are C22:5 and C22:6, which make up the bulk of the other acids. However, many other fatty acids of various levels of unsaturation are also present.

Biodiesel Issues

Unknown.

Agricultural Issues

There is extensive research work in Australia and overseas on maximising the food benefits of fish oils, especially in fish-farming operations.

Hemp[18]

---

[18] Wikipedia, *Picture Author Durbetorte*, downloaded July 24, 2024.

Scientific Name: *Cannabis sativa*, subspecies sativa

This is the form of hemp grown in Europe and Canada for industrial purposes.

Fatty Acid Composition

	C16:0	**C18:0**	**C18:1**	**C18:2**	**C18:3**	**C22:1**	**Total**
Hemp	7.5	2.5	13	60	23.5	0	106.5

Estimated Biodiesel Properties

	S.N.	**I.N.**	**C.N.**	**Density**	**Viscosity**	**%D(360)**
Estimate	214.5	185.0	30.1	886	4.26	100
Literature		157.5				

Biodiesel Issues

The fatty acids have a high degree of unsaturation as evidenced by the high iodine number and C18:3 content. Biodiesel produced from the hemp FAME would be expected to be less stable than soybean and would require a higher level of antioxidants. The cetane number is low. Potentially, this could be used blended with less saturated feedstock.

Agricultural Issues

Hemp cultivation of cultivars (*sativa*) low in tetrahydrocannabinol (THC) (as opposed to the *C. indica* variety) is permitted in many jurisdictions. Hemp fibre has many industrial uses.

Honge[19] *(Pongamia pinnata)*

Scientific Name: *Pongamia pinnata*
Other Names: Karanj, pongamia

Fatty Acid Composition

	C16:0	C18:0	C18:1	C18:2	C18:3	C22:1	Total
Honge	14	18	55	9	0	0	96

Estimated Biodiesel Properties

	S.N.	I.N.	C.N.	Density	Viscosity	%D(360)
Estimate	201.8	65.9	58.5	889	4.35	98
Literature						

Biodiesel Issues

The FAME distribution from honge is likely to be similar to Canola and tallow. It has a high oleic acid content. A good-quality biodiesel would be expected.

---

[19] Wikipedia, *Picture Author Forest and Kim Starr*, downloaded July 24, 2024.

Agricultural Issues

P. pinnata is a legume, medium-sized evergreen tree with a spreading crown and short bole. It is often grown as an ornamental. The tree is nitrogen fixing. It is often cultivated along roads and canal banks. The annual rainfall required is 500 to 2,500 mm in its natural habitat. *Pongamia* produces root suckers and is generally unsuitable for agroforestry.

The tree is naturally distributed in tropical and temperate Asia. It is well adapted to arid zones.

Jatropha[20]

Scientific Name: *Jatropha curcas*

There are about 170 subspecies of *Jatropha*. *J. curcas* is the type being promoted for biodiesel production.

Fatty Acid Composition

	C16:0	**C18:0**	**C18:1**	**C18:2**	**C18:3**	**C22:1**	**Total**
Jatropha	4.2	6.9	43.1	34.3	1.4	0	89.9

---

[20] Wikipedia, *Picture Author Anish Nellickal*, downloaded July 24, 2024.

Estimated Biodiesel Properties

	S.N.	I.N.	C.N.	Density	Viscosity	%D(360)
Estimate	179.8	104.9	53.1	880	3.34	89.9
Literature		101.7				

Biodiesel Issues

The FAME distribution of *Jatropha* oil lies somewhere between Canola and palm olein. The product biodiesel would thus be expected to be of good quality. It is widely promoted as a source for biodiesel in India.

Agricultural Issues

Jatropha is a small tree or shrub. Jatropha is resistant to droughts and pests. Some of the seed oil contents are poisonous, and the oil is not suitable for human consumption. The leaves have insecticidal properties. In some jurisdictions, it is considered a noxious weed.

Kapok[21]

---

[21] Wikipedia, *Picture Author Alexandre*, downloaded July 24, 2024.

Scientific Name: *Ceiba pentandra*
Also classified as *Eriodendron anfractuosum*

Fatty Acid Composition

	C16:0	**C18:0**	**C18:1**	**C18:2**	**C18:3**	**C22:1**	**Total**
Kapok	22	3	21	37	0	0	83

In Kapok, there is a significant amount of cyclopropenoid fatty acids, typically 14% of the fatty acids.

Estimated Biodiesel Properties

	S.N.	**I.N.**	**C.N.**	**Density**	**Viscosity**	**%D(360)**
Estimate	169.7	86.0	59.1	875	2.84	100
Literature		98.0				

Biodiesel Issues

The iodine and cetane values indicate that kapok could potentially produce a good-quality biodiesel. However, the estimated viscosity is low relative to the standard. The cyclopropenoid fatty acids are very reactive and are said to decompose on extraction and work-up to produce the oil product. The nature and fate of the decomposition products are unknown, and hence, there is some uncertainty to the final FAME distribution in a biodiesel product. If the breakdown of the cyclopropenoids results in shorter-chain fatty acids, this could compromise the final fuel quality. From the data available, the product quality cannot be predicted.

Agricultural Issues

Kapok is a large tree (height 60–70 m). The trunk and branches are covered in thorns. Pods contain seeds surrounded by a water-resistant fibre. Collection is labour-intensive. The tree is cultivated in Southeast Asia.

Mahua[22]

Scientific Name: *Madhuca longifolia*
Other Names: Honey tree, butter tree

Fatty Acid Composition

	C16:0	C18:0	C18:1	C18:2	C18:3	C22:1	Total
Mahua	24.5	22.7	37	14.3	0	0	98.5

Estimated Biodiesel Properties

	S.N.	I.N.	C.N.	Density	Viscosity	%D(360)
Estimate	200.4	59.3	60.2	867	4.89	98.5
Literature		62.5				

Biodiesel Issues

As evidenced by the low iodine number, the oil contains a high proportion of saturated fatty acid. There is a large portion of C16:0. The distribution is similar to that for tallow but with a higher level of unsaturated acids at C18:2. This should improve its cold flow properties relative to tallow.

---

[22]Wikipedia, *Picture Author Nvvchar*, downloaded July 24, 2024.

Agricultural Issues

This is an evergreen or semi-evergreen tree with a spreading crown. The tree is found in North Indian forests and plains. It is cultivated for its seed and wood. It produces between 20 and 200 kg seeds per year.

Milk Bush[23]

Scientific Name: *Cascabela thevetia* and *Thevetia peruviana*
Other Names: Yellow oleander, lucky nut

Fatty Acid Composition

	C16:0	**C18:0**	**C18:1**	**C18:2**	**C18:3**	**C22:1**	**Total**
Milk Bush	20.12	7.39	47.47	14.77	0.47	0	90.22

---

[23] Wikipedia, *Picture Authors Anshik Kumar Tiwari and Ji-Elle*, downloaded July 24, 2024.

Estimated Biodiesel Properties

	S.N.	I.N.	C.N.	Density	Viscosity	%D(360)
Estimate	189.6	71.3	59.1	874	3.53	90.9
Literature		80.0				

Biodiesel Issues

The FAME distribution is within the bounds of the principal feedstocks and should be suitable for biodiesel.

Agricultural Issues

It is a poisonous shrub native to Mexico and Central America. Sap and fruit are toxic. It is an evergreen tropical shrub that bears yellow, trumpet-like flowers, and its fruit is deep red/black in colour encasing a large seed-bearing resemblance to a Chinese "lucky nut" (Wikipedia).

Mustard[24]

Scientific Name: *Brassica nigra*

---

[24] Wikipedia, *Picture Author Nafiur Rahman*, downloaded July 24, 2024.

Mustards are several plant species in the genera *Brassica* and *Sinapis* whose small mustard seeds are used as a spice and, by grinding and mixing them with water, vinegar or other liquids, are turned into a condiment also known as mustard. The seeds are also pressed to make mustard oil, and the edible leaves can be eaten as mustard greens.

Fatty Acid Composition

	C16:0	C18:0	C18:1	C18:2	C18:3	C22:1	Total
Mustard	2.6	0.8	23.2	8.9	10.4	43.1	89

Estimated Biodiesel Properties

	S.N.	I.N.	C.N.	Density	Viscosity	%D(360)
Estimate	179.6	104.7	53.1	876	4.90	46.0
Literature		101.0				

Biodiesel Issues

The *Brassica* genus also contains rapeseed, and the mustard oil is very similar in character. The FAME would be expected to have similar properties. There is a high erucic acid content, which will cause an issue with the T(90) distillation point. However, in hydrogenation processes to produce green diesel, there can potentially be a high yield of SAF.

Agricultural Issues

A wide range of crops and wild (weed) varieties are available. It is a well-known agricultural crop similar to rape and Canola.

Neem[25]

Scientific Name: *Azadirachta indica*
Other Names: Margosa, nimtree, Indian lilac

Fatty Acid Composition

	C16:0	C18:0	C18:1	C18:2	C18:3	C22:1	Total
Neem	14	18	55	9	0	0	96

Estimated Biodiesel Properties

	S.N.	I.N.	C.N.	Density	Viscosity	%D(360)
Estimate	201.8	65.9	58.5	889	4.35	98.0
Literature		72.5	57.8			

Biodiesel Issues

Neem oil is similar in composition to Mahua (see above). As evidenced by the iodine number, the oil contains a high proportion of saturated fatty acids. There is a large portion of C16:0. The distribution is similar to that

---

[25] Wikipedia, *Picture Author Banksboomer*, downloaded July 24, 2024.

for tallow but with a higher level of unsaturated acids at C18:2. This should improve its cold flow properties relative to tallow.

Agricultural Issues

Neem is a fast-growing tree that can reach a height of 15 to 20 m, rarely to 35 to 40 m. It is evergreen, but under severe drought, it may shed most or nearly all of its leaves. The branches are widespread. The fairly dense crown is roundish or oval and may reach the diameter of 15 to 20 m in old, free-standing specimens.

The neem tree is noted for its drought resistance. Normally, it thrives in areas with sub-arid to sub-humid conditions, with an annual rainfall between 400 and 1,200 mm. It can grow in regions with an annual rainfall below 400 mm, but in such cases, it depends largely on the groundwater levels. Neem can grow in many different types of soil, but it thrives best on well-drained deep and sandy soils (pH 6.2–7.0). It is a typical tropical/subtropical tree and exists at annual mean temperatures between 21°C and 32°C. It can tolerate high to very high temperatures. It does not tolerate temperature below 4°C (leaf shedding and death may ensue).

Palm Oil[26]

Scientific Name: *Elaeis guineensis; E. oleifera*

Fatty Acid Composition

The palm fruit is the source of both palm oil (extracted from palm fruit) and palm kernel oil (extracted from the fruit seeds). Palm olein is the

---

[26] Wikipedia, *Picture Author Marco Schmidt and Bongoman Respectively*, downloaded July 26, 2024.

liquid fraction obtained by fractionation of palm oil after crystallisation at controlled temperatures. The physical characteristics of palm olein differ from those of palm oil. It is fully liquid in warm climate and has a narrow range of glycerides. Palm stearin is the more solid fraction obtained by fractionation of palm oil after crystallisation at controlled temperatures. It is thus a co-product of palm olein. It is traded at a discount to palm oil and palm olein, making it a cost-effective ingredient in several applications.

	C16:0	**C18:0**	**C18:1**	**C18:2**	**C18:3**	**C22:1**	**Total**
Palm kernel	8.25	2.15	15.5	2.25	0	0	28.15
Palm stearin	44.2	4.4	39.6	10.25	0.25	0	98.7
Palm olein	54.2	4.95	29.15	6.7	0.1	0	95.1
Palm oil	40.15	4.15	43.3	11	0.2	0	98.8

Palm kernel FAME has a large amount of C12 and C14 fatty acids and is similar in properties to coconut and would be expected to have similar properties.

Estimated Biodiesel Properties

Palm kernel	**S.N.**	**I.N.**	**C.N.**	**Density**	**Viscosity**	**%D(360)**
Estimate	260.4	18.0	63.2	871	2.77	100
Literature		18.0				
Palm stearin						
Estimate	203.5	38.8	64.4	875	4.04	96.5
Literature		30				
Palm olein						
Estimate	207.6	59.6	59.2	877	4.36	100
Literature		56 (min)				

Biodiesel Issues

The major portion of the fatty acids in palm kernel oil lies below C16. This oil is very similar to coconut with the same undesirable properties for biodiesel namely too low boiling point. However, this makes it a good feedstock for aviation kerosene. The other oils from the palm tree are suitable for biodiesel production with the fatty acid distribution in the diesel range. The preferred feedstock is palm olein, which is lighter and has better cold flow properties than palm stearin.

Of the vegetable oils, its mass cultivation results in a relatively low price, which makes the oil and its derivatives attractive for the mass production of biofuels in biorefineries. The Neste Oil biorefinery in Singapore has a capacity of 899,000t/y.

Agricultural Issues

Palm oil is a majorly traded commodity. Over 35% of the world's consumption of vegetable oils is palm oil. Palm plantations dominate the landscape of much of Southeast Asia, which as a monoculture significantly reduces biodiversity. Palm oil is grown in the tropics, which facilitates typically three crops per year; hence, palm oil has a very high production rate per hectare and is hence an attractive crop for the replacement of tropical rain forests and consequential environmental damage. Palm oil production is very large with the main products being soaps and edible oils and margarines. There is a large trade that is well reported.

Peanut[27]

Scientific Name: *Arachis hypogaea*
Other Names: Groundnut. Thousands of cultivars.

---

[27] Wikipedia, *Picture Author Abhay Iari*, downloaded July 26, 2024.

Fatty Acid Composition

	C16:0	C18:0	C18:1	C18:2	C18:3	C22:1	Total
Peanut	10.4	8.9	47.1	32.9	0.5	0.2	100

Estimated Biodiesel Properties

	S.N.	I.N.	C.N.	Density	Viscosity	%D(360)
Estimate	201.0	103.6	50.1	877	4.27	99.8
Literature		93.0				

Biodiesel Issues

The fatty acid composition indicates that the FAME composition would be partway between soybean and palm olein. This indicates that peanut oil would make a good-quality biodiesel.

Agricultural Issues

The flower of the *A. hypogaea* is borne above ground, and after it withers, the stalk elongates, bends down and forces the ovary underground. When the seed is mature, the inner lining of the pods (called the seed coat) changes colour from white to a reddish brown. The entire plant, including most of the roots, is removed from the soil during harvesting. Peanuts are widely grown, and the nuts are widely traded.

Petroleum Nut[28]

Scientific Name: *Pittosporum resiniferum*
Other Names: Resin cheesewood

Fatty Acid Composition

	C16:0	**C18:0**	**C18:1**	**C18:2**	**C18:3**	**C22:1**	**Total**
Petroleum nut	10	3	25	52	10		100

Estimated Biodiesel Properties

	S.N.	**I.N.**	**C.N.**	**Density**	**Viscosity**	**%D(360)**
Estimate	201.6	144.3	40.9	882	3.9	100
Literature[29]			52	865	4.3	

---

[28] Wikipedia, *Picture Author Dick Culbert*, downloaded July 26, 2024.

[29] R. Kipkoech and M. Takase, "Production of Biodiesel from Crude *Pittosporum Resiniferum* Oil Using Heterogeneous Solid Base Catalyst," *Nano Select* **5**, e202300132 (2024). Doi:10.1002/nano.202300132.

Biodiesel Issues

The high level of unsaturated fatty acids may lead to relatively poor oxidative stability of the FAME. This seems to be borne out by the literature. The fruits burn brightly when ignited. The oil from the fruit contains dihydroterpene ($C_{10}H_{18}$) and considerable quantities of n-heptane.

Agricultural Issues

The petroleum nut is a tree that grows in the Philippines and Malaysia. It derives its name from the odour of the fruit, being similar to petroleum fuels. The fruit is harvested from the wild.

Poppy[30]

Scientific Name: *Papaver somniferum*
　　The variety of interest is the opium poppy because of proposals to develop a poppy seed industry for biodiesel in Afghanistan (as opposed to selling heroin) and as a by-product from the cultivation of the opium poppy for pharmaceutical purposes.

Fatty Acid Composition

	C16:0	C18:0	C18:1	C18:2	C18:3	C22:1	Total
Poppy	10	2	11	72	5	0	100

Estimated Biodiesel Properties

	S.N.	**I.N.**	**C.N.**	**Density**	**Viscosity**	**%D(360)**
Estimate	201.7	154.2	38.7	884	3.80	100
Literature		133.4				

Biodiesel Issues

The fatty acid distribution is similar to that of soybean. Olefin contents are high as witnessed by the high iodine number and low cetane.

Agricultural Issues

The cultivation of *P. somniferum* is banned or highly regulated in most jurisdictions. It is an annual herb growing about 100 cm tall.

Prickly Acacia[31]

Scientific Name: *Acacia nilotica, Vachellia nilotica*

---

[31] Wikipedia, Wikimedia Foundation, *Picture Author Harvinder Chandigarh*, downloaded July 29, 2024.

Fatty Acid Composition

	C16:0	C18:0	C18:1	C18:2	C18:3	C22:1	Total
Prickly acacia	18.1	7.8	28.1	38.2	2.62	0	94.82

Estimated Biodiesel Properties

	S.N.	I.N.	C.N.	Density	Viscosity	%D(360)
Estimate	195.2	101.8	51.4	877	3.70	95.0
Literature						

Biodiesel Issues

The fatty acid distribution is in the range of the principal feedstocks, and therefore, the FAME has the potential to be a good-quality biodiesel.

Agricultural Issues

A. nilotica (Thorn mimosa) is a species of Acacia (wattle) native to Africa and the Indian subcontinent. It grows to 20 m with a dense spheric crown. It is also currently an invasive species of significant concern in Australia and the United States. The extrudate from the tree is known as *gum arabic*.

Radish[32]

---

[32] Wikipedia, Wikimedia Foundation, *Picture Author Alex Lockton*, downloaded July 29, 2024.

Scientific Name: *Raphanus raphanistrum*
Other Names: Wild radish; *R. sativus* is the cultivated root vegetable.

Fatty Acid Composition

	C16:0	C18:0	C18:1	C18:2	C18:3	C22:1	Total
Radish	5	2	18	11	13	37	86

Estimated Biodiesel Properties

	S.N.	I.N.	C.N.	Density	Viscosity	%D(360)
Estimate	181.6	107.8	52.1	877	4.59	49.0
Literature		108.0				

Biodiesel Issues

Of the same family is rapeseed, and hence, it has a similar fatty acid distribution (see also mustard). The FAME should give a biodiesel of similar quality to rapeseed oil. There will be an issue in achieving the distillation T(90) point.

Agricultural Issues

It is native to North Africa, Europe and parts of Western Asia. Wild radish is a noxious weed over much of temperate Australia.

Wild radish is an erect annual to 1 m high, which reproduces by seed. When crushed, the basal leaves have a strong turnip-like smell. Stems are bluish green, often red at the base. Flowers have four petals, are white, yellow or pink to mauve and are stalked. Seed pods are 2 to 9 cm long, ribbed and distinctly constricted between the seeds. At maturity, during crop harvest, the pods break up readily at the constriction points into segments that contain one seed.[33]

---

[33] Australia: Weed Cooperative Research Centre.

Ramtil[34]

Scientific Name: *Guizotia abyssinica*
Other Names: Niger, Noog, Nug, Kalatil, Karala (India)

Fatty Acid Composition

	C16:0	C18:0	C18:1	C18:2	C18:3	C22:1	Total
Ramtil	8.5	7.9	27.0	55.0	0.0	0.0	98.4

Estimated Biodiesel Properties

	S.N.	I.N.	C.N.	Density	Viscosity	%D(360)
Estimate	198.6	124.1	45.9	880	3.92	98.4
Literature		112				

Biodiesel Issues

The fatty acid distribution is within the bounds of the principal feedstock. With the low iodine value, this indicates that the FAME will produce a good biodiesel.

---

[34] Wikipedia, *Picture Author Silvestresbrasileiros*, downloaded July 29, 2024.

Agricultural Issues

Native to Ethiopia, Eritrea and Malawi and grown in India. As an erect, stout, branched annual herb, grown for its edible oil and seed. Its cultivation originated in the Ethiopian highlands and has spread to other parts of Ethiopia (Wikipedia).

It has potential for weed-like properties. Seeds exported from Ethiopia and India are often sterilised to prevent growth.

Rapeseed[35]

Scientific Name: *Brassica napus*
Other Names: Rapeseed (*B. napus*), also known as rape, oilseed rape, rapa, rapeseed (in the case of one particular group of cultivars), Canola (see above).

Fatty Acid Composition

	C16:0	C18:0	C18:1	C18:2	C18:3	C22:1	Total
Rapeseed	2.7	2.8	21.9	13.1	8.6	50.9	100

---

[35] Wikipedia, *Picture Author Musem de Toulouse*, downloaded July 29, 2024.

Estimated Biodiesel Properties

	S.N.	I.N.	C.N.	Density	Viscosity	%D(360)
Estimate	182.8	105.3	52.5	876	5.38	49.1
Literature		103.0		880	5.73	

Biodiesel Issues

Rapeseed is a principal feedstock for the production of biodiesel in Europe. Rapeseed FAME has a high viscosity, above the regulated limit. The high erucic acid content causes an issue with achieving the distillation T(90) values.

Agricultural Issues

Rapeseed is a bright yellow flowering member of the family *Brassicaceae* (mustard or cabbage family).

Rapeseed is grown for the production of animal feed, vegetable oil for human consumption, and biodiesel; leading producers include the European Union, Canada, the United States, Australia, China and India.

Rice Bran

Scientific Name: *Oryza sativa* and *O. glaberrima*

Fatty Acid Composition

	C16:0	C18:0	C18:1	C18:2	C18:3	C22:1	Total
Rice bran	16.4	1.8	42.3	37.1	1.3	0	98.9

Estimated Biodiesel Properties

	S.N.	I.N.	C.N.	Density	Viscosity	%D(360)
Estimate	202.5	109.2	48.7	881	4.09	99.4
Literature		104.5				

Biodiesel Issues

The fatty acid distribution falls within the range of the principal feed-stocks; this would be expected to give a good quality biodiesel.

Agricultural Issues

Rice is a major world-traded commodity and grown extensively. Rice bran oil is the oil extracted from the germ and inner husk of rice. It is notable for its very high smoke point of 490°F (254°C) and its mild flavour, making it suitable for high-temperature cooking methods such as stir-frying and deep-frying. It is popular as a cooking oil in several Asian countries, including Japan and China.

Rubber Seed[36]

Scientific Name: *Hevea brasiliensis*

Fatty Acid Composition

	C16:0	**C18:0**	**C18:1**	**C18:2**	**C18:3**	**C22:1**	**Total**
Rubber	10.2	8.7	24.6	39.6	16.3	0	99.4

---

[36] Wikipedia, *Picture Author Vyacheslav Argenberg*, downloaded July 29, 2024.

Estimated Biodiesel Properties

	S.N.	I.N.	C.N.	Density	Viscosity	%D(360)
Estimate	200.4	138.7	42.3	880	3.91	99.4
Literature		136.2				

Biodiesel Issues

The oil is produced from the seed of the para rubber tree, which is widely cultivated in Southeast Asia for the production of natural rubber.

The high level of the highly unsaturated C18:3, verified by the high iodine number, indicates that FAME produced from the oil would require treatment with antioxidants. The biodiesel could find a role as a blend with other less unsaturated feedstocks.

Agricultural Issues

Plantations are found across Southeast Asia.

Safflower[37]

Scientific Name: *Carthamus tinctorius*

---

[37] Wikipedia, *Picture is in Public Domain*, downloaded July 29, 2024.

Fatty Acid Composition

	C16:0	**C18:0**	**C18:1**	**C18:2**	**C18:3**	**C22:1**	**Total**
HL Safflower	6.6	3.3	14.4	75.5	0.1	0	99.9
HO Safflower	5.7	2.3	73.6	15.8	0	0	97.4

Biodiesel Properties

HL Safflower	**S.N.**	**I.N.**	**C.N.**	**Density**	**Viscosity**	**%D(360)**
Estimate	201.0	150.2	39.7	885	3.86	100
HO Safflower						
Estimate	196.8	95.0	52.7	879	4.11	97.5
Literature		95.0				

Biodiesel Issues

There are two types of safflower that produce different kinds of oil: one high in monounsaturated fatty acid (C18:1, oleic acid) and the other high in polyunsaturated fatty acid (C18:2, linoleic acid). Currently, the predominant oil market is for the former, which is lower in saturates than olive oil. Two fatty acid compositions are provided, one for high oleic and the other for high linoleic acid. High linoleic FAME has properties outside the range of the principal feedstocks. The high oleic FAME has good properties for the production of biodiesel.

Agricultural Issues

Safflower (*C. tinctorius*) is a highly branched, herbaceous, thistle-like annual, usually with many long sharp spines on the leaves. Plants are 30 to 150 cm tall with globular flower heads (capitula) and commonly, brilliant yellow, orange or red flowers that bloom in July (northern hemisphere). Each branch will usually have from one to five flower heads containing 15 to 20 seeds per head. Safflower has a strong taproot, which enables it to thrive in dry climates, but the plant is very susceptible to frost injury from stem elongation to maturity.

Soybean[38]

Scientific Name: *Glycine L. max*

Fatty Acid Composition

	C16:0	C18:0	C18:1	C18:2	C18:3	C22:1	Total
Soybean	10.3	4.7	22.5	54.1	8.3	0	99.9

Estimated Biodiesel Properties

	S.N.	I.N.	C.N.	Density	Viscosity	%D(360)
Estimate	201.6	141.2	41.6	882	3.92	100
Literature		140.0	57	886	4.28	

Biodiesel Issues

Soybean is the major feedstock for the production of biodiesel in the USA. It has a higher olefin content than rapeseed and, as a consequence, produces FAME, which are more susceptible to oxidation.

Agricultural Issues

The soybean (United States) or soya bean (UK) (*Glycine max*) is a species of legume native to East Asia. It is an annual plant that may vary in

---

[38] Wikipedia, *Picture Author H. Zell*, downloaded July 29, 2024.

growth, habit and height. It may grow prostrate, not growing higher than 20 cm (7.8 inch), or even up to 2 m (6.5 ft) in height. The pods, stems and leaves are covered with fine brown or grey pubescence.

Soybeans are an important global crop, providing oil and protein. The bulk of the crop is solvent-extracted for vegetable oil, and then defatted soy meal is used for animal feed. A small proportion of the crop is consumed directly by humans. Soybean products appear in a large variety of processed foods.

Sunflower[39]

Scientific Name: *Helianthus annuus*

Fatty Acid Composition

	C16:0	C18:0	C18:1	C18:2	C18:3	C22:1	Total
Sunflower	6.6	4.6	26.7	61.2	0.1	0.1	99.3

Estimated Biodiesel Properties

	S.N.	I.N.	C.N.	Density	Viscosity	%D(360)
Estimate	202.5	135.8	42.7	882	3.94	99.4
Literature		130.5			4.24	

---

[39] Wikipedia, *Picture Author Shirleybolling2005*, downloaded July 29, 2024.

Biodiesel Issues

Sunflower oil contains a large amount of the unsaturated C18:2 fatty acid. This contributes to a relatively high iodine number. However, the oil contains less of the highly unsaturated C18:3 fatty acid. There should be no issues with biodiesel made from this feedstock from the standpoint of FAME composition.

Agricultural Issues

The sunflower (*H. annuus*) is an annual plant native to the Americas in the family *Asteraceae*, with a large flowering head (inflorescence). The stem of the flower can grow as high as 3 m tall, with the flower head reaching up to 30 cm in diameter with the "large" seeds. The term "sunflower" is also used to refer to all plants of the genus *Helianthus*, many of which are perennial plants.

Tallow

Tallow is the collective name for animal fats, usually beef or lamb.

Fatty Acid Composition

	C16:0	C18:0	C18:1	C18:2	C18:3	C22:1	Total
Tallow beef	25.2	19.2	48.9	2.7	0.5	0	96.5
Tallow lamb	24.6	30.5	36.0	4.3	0	0	95.4
Tallow pork	25.5	15.8	47.1	8.9	1.1	0	98.4

Biodiesel Properties

Beef	S.N.	I.N.	C.N.	Density	Viscosity	%D(360)
Estimate	205.2	50.3	61.6	898	4.60	100
Literature		41.5			4.50	
Lamb						
Estimate	205.3	40.2	63.8	882	4.71	100
Literature		41.2				
Pork (lard)						
Estimate	204.5	61.6	59.1	882	4.52	100
Literature		58.0			4.54	

Biodiesel Issues

Compared to vegetable oils, tallow fatty acids are characterised by a high level of saturated fats (C16:0 and C18:0). This leads to a highly stable biodiesel, which has a high melting point (poor cold flow properties).

Agricultural Issues

Tallow is widely available from the large meat processing industries of Australia, New Zealand and the United States and at generally lower cost than vegetable oils. Uptake of tallow is limited by the cold flow properties of the final biodiesel. Tallow is usually used along with more expensive vegetable oils to attain a practical outcome.

Tung[40]

Scientific Name: *Aleurites fordii*
Other Names: *Vernicia fordii*. There are a number of cultivars.

---

[40] Wikipedia, *Picture Author KENPEI*, downloaded July 29, 2024.

Fatty Acid Composition

	C16:0	**C18:0**	**C18:1**	**C18:2**	**C18:3**	**C22:1**	**Total**
Tung	3.1	2.4	11.2	14.6	69	0	100.3

Estimated Biodiesel Properties

	S.N.	**I.N.**	**C.N.**	**Density**	**Viscosity**	**%D(360)**
Estimate	201.9	225.7	22.5	890	3.43	100
Literature		167.5				

Biodiesel Issues

Tung oil contains large amounts of the highly unsaturated C18:3 fatty acid. The presence of C18:3 acid in these high concentrations would be expected to result in oxidative instability of the FAME. It is similar in character to linseed oil and would normally not be considered suitable for biodiesel production.

Agricultural Issues

V. fordii (Tung Tree; syn. *A. fordii*) is a species of *Vernicia* in the spurge family, native to southern China, Burma and northern Vietnam. It is a small- to medium-sized deciduous tree growing to 20 m tall, with a spreading crown. It is classed as an invasive species in Florida.

The tree is listed as a poisonous plant.

Abbreviations

AVTUR	Aviation turbine fuel — Jet fuel
a	annum (year)
bbl	petroleum barrel
bbl/d	barrels per day
bcfd	billions of cubic feet per day

(Continued)

BFW	Boiler feed water (high purity water)
BTU	British Thermal Unit
BTX	benzene, toluene and xylene mixture
C	Degrees Centigrade (Celsius)
Capex	Capital cost
cf	cubic foot
CIF	container, insurance and freight (destination port price)
cm	cubic metre
CNG	Compressed natural gas
DME	Dimethyl ether
F	Degrees Fahrenheit
FAME	Fatty acid methyl ester
FOB	free on board (embarkation port price)
GJ	gigajoule
HDPE	High-density polyethylene
HHV	higher heating value (gross)
HP	horsepower
K	Degrees Kelvin (absolute temperature scale)
K-factor	UOP or Watson paraffinicity factor
kt/y	thousand metric tonnes per year
kW	kilowatt
kWh	kilowatt-hour
L	litre
lb	pound
LDPE	Low-density polyethylene
LHV	lower heating value (net)
LLDPE	linear low-density polyethylene
LNG	Liquefied natural gas (mainly methane)
LPG	Liquefied petroleum gas (usually propane and butane)
MCH	Methyl-*cyclo*-Hexane
MM$	million US dollars (2007)
MMBTU	million (US Customary) BTU

(Continued)

(*Continued*)

Mt	million metric tonnes
MTC	Methanol to chemicals
MTG	Methanol to gasoline (Mobil process)
MTO	Methanol to olefins
NGL	Natural gas liquids (ethane, propane, butane)
OPEX	Non-feedstock operating costs
PJ	Peta joule ($10e^{15}$ joules)
PM	Particulate matter (exhaust)
PONA	paraffins, olefins, naphthenes and aromatics
PP	polypropylene
R	Degrees Rankin (absolute temperature scale in °F)
ROC	Return on capital
S.I.	Système International d'Unités; metric units
SAK	Synthetic aromatic kerosene
SAPO	Silica, alumina, phosphorus, oxygen — molecular sieve
SMR	Steam methane reformer
t	metric tonne
t/y	metric tonnes per year
VGO	vacuum gas oil
WC	Working capital

INDEX